Challenging

—— the ——

Unchallengeable

Einstein's Theory of Special Relativity

CHALLENGING

—— THE ——

UNCHALLENGEABLE

Einstein's Theory of Special Relativity

John D. Frey

Email
relativity880@gmail.com

Archway Publishing books may be ordered through booksellers or by contacting:

Archway Publishing
1663 Liberty Drive
Bloomington, IN 47403
www.archwaypublishing.com
1 (888) 242-5904

ISBN: 978-1-4808-5618-9 (sc)
ISBN: 978-1-4808-5619-6 (e)

Library of Congress Control Number: 2017919427

Print information available on the last page.

Archway Publishing rev. date: 1/30/2018

Contents

INTRODUCTION

When it comes to the principles and logic upon which special relativity is built, the scientific community seems to accept their validity as either certain or as certain as the foundational concepts of any theory can be. There are various reasons for the firm acceptance of these principles and logic, and among them is the fact that they have existed and been accepted by the scientific community for more than 100 years. A second reason is that various experiments have tested the principles, logic, and results of special relativity and have been used in their support. A third reason centers on the brilliance of the person who created special relativity, Albert Einstein. Hence, we find an almost universal conviction in the world of science that the core concepts upon which special relativity is built are beyond serious challenge. Since the purpose of this book is to test and challenge these core concepts, the title of the book speaks about "challenging the unchallengeable."

As the author of this book, I will describe myself as a nonprofessional lover of science who has spent more than ten years as an ardent researcher into every aspect of special relativity. This research not only involved an enormous amount of reading and reflecting about special relativity, but it especially involved an ongoing dialogue with a number of awesome professors of physics who were willing to share their time and expertise with me. Over the years I was able to place before them whatever concepts, questions, or experimental procedures that I was considering, and they gave me their reactions. This, in turn, led to hundreds of email exchanges and, at times, to some very intense dialogue. On some occasions their responses helped me to see the error of my ways. On other occasions their responses helped in firming up my own divergent viewpoints. But behind my conclusions about special relativity, there was input from some caring and learned professors, and I shall forever be grateful to them for their willingness to share their knowledge and time with me. However, I am not posting their names, because some might not want to be listed in a book that seeks to challenge the core concepts of special relativity.

In sending this book out into the world, I am making no claims about it being the last word on the subject of special relativity. I do think, however, that it presents innovative concepts and step by step logic that are worthy of consideration. It is my hope that by being written in everyday, nontechnical language, this book will allow people from many walks of life to participate in a vital discussion about the makeup of the universe in which we live.

Overview of Special Relativity

In June 1905, Albert Einstein submitted an article to the German publication *Annalen der Physik* entitled "On the Electrodynamics of Moving Bodies." Within its pages, he presented two principles and a logical application of these principles that led to two unique conclusions when systems are in uniform motion with respect to each other - the dilation or slowing of time and the contraction of a system's length. In September of that same year, Einstein submitted a short, three-page article to *Annalen der Physik*, "Does the Inertia of a Body Depend Upon Its Energy Content?" which concerned the equivalence of energy and mass. It contained, in essence, the most famous equation in the world, $E=mc^2$, as well as a related equation $E=mc^2/\sqrt{(1-v^2/c^2)}$. Eventually, the contents of these two articles and their dramatic conclusions came to be known as the *special theory of relativity*.

Einstein reiterated the contents of the two articles and offered further explanations of them in a number of other published works, such as "On the Relativity Principle and the Conclusions Drawn from It, "a document of 1907; "The Theory of Relativity," a 1911 lecture given in Zurich; *Relativity: The Special and the General Theory*, a book offered in 1916; and *The Meaning of Relativity*, a book published by Princeton University containing four lectures presented by Einstein at Princeton in 1921.

As to the contents of this book, chapter one presents the two principles upon which the time dilation and length contraction of special relativity are built. The second chapter offers concepts that were created before special relativity and were preludes to the time dilation and length contraction of special relativity. Chapter 3 gives an overview concerning the way in which the claims of special relativity bring about time dilation and length contraction through the functioning of its two principles.

Chapters 4 through 8 present various experiments and concepts that test the validity of special relativity's claims and reveal how and why these claims fail this testing. In chapters

9 through 13, it is shown why prominent experiments and events that are used to support the claims of special relativity fail to accomplish this purpose. Chapter 14 considers the claim of special relativity that is most likely to have caused special relativity to fail its testing in chapters 4 through 8.

Thought Experiments and Amazing Human Abilities

In explaining special relativity, Einstein frequently used imaginary experiments that are conducted with concepts and words rather than with concrete objects and actions. This practice is understandable since in today's world there are often no physical experiments that can reveal what takes place in accord with the claims of special relativity. It is possible, for instance, to imagine spaceships traveling near the speed of light and make logical deductions from such events even though there are no man-made vessels that can come anywhere near such speeds. In this paper we will follow Einstein's example by using verbal experiments as he used them.

English System of Measurements

The English system of measurements will be used throughout the book. The reason comes from the very convenient fact that the constant speed of light is approximately one-foot-per-nanosecond. This will be the speed that is used for light in all the experiments that are presented in this book. The actual speed of light is .983571-foot-per-nanosecond.

CHAPTER ONE

Special Relativity's Two Principles

As shown in chapter 2 of Einstein's "On the Electrodynamics of Moving Bodies," the time dilation and length contraction of special relativity are built on two principles - the relativity principle and the constant speed of light principle. In this chapter, we present an overview of both principles.

The Relativity Principle

The relativity principle of special relativity has two basic aspects. Concerning the first aspect, the principle insists that when there is uniform motion between two systems, A and B, it is completely the motion of A with respect to B and completely the motion of B with respect to A. In chapter 18 of *Relativity: The Special and the General Theory*, Einstein describes this facet of the relativity principle with these words:

> The basal principle, which was the pivot of all our previous considerations, was the *special* principle of relativity, *i.e.* the principle of the physical relativity of all uniform motion. Let

us once more analyze its meaning carefully. It was at all times clear that, from the point of view of the idea it conveys to us, every motion must be considered only as a relative motion. Returning to the illustration we have frequently used of the embankment and the railway carriage, we can express the fact of the motion here taking place in the following two forms, both of which are equally justifiable: (*a*) The carriage is in motion relative to the embankment.(*b*) The embankment is in motion relative to the carriage.

Hence, as shown by this statement, the relativity principle of special relativity deals with the uniform motion between systems such as between a train and its embankment. The principle insists that this uniform motion is the motion of the train relative to the embankment and observers on the embankment and the motion of the embankment relative to the train and observers in the train.

In the first paragraph of his 1911 paper, "The Theory of Relativity," Einstein presents a second aspect of the relativity principle with the following account:

Picture to yourself two physicists. Let both physicists be equipped with every physical instrument imaginable; let each of them have a laboratory. Suppose that the laboratory of one of the physicists is arranged somewhere in an open field, and that of the second in a railroad car traveling at constant velocity in a given direction. The principle of relativity states the following: if, using all their equipment, these two physicists were to study all the laws of nature, one in his stationary laboratory and the other in his laboratory on the train, they would discover exactly the same laws of nature, provided that the train is not shaking and is traveling in uniform motion. Somewhat more abstractly, we can say: according to the principle of relativity, the laws of nature are independent of the translational motion of the reference system.

Hence, according to Einstein's relativity principle, if there is uniform motion between system A and system B after the two systems had been at rest with each other, this motion can cause no changes in the way an event happens in A from the way the event happened in A before this motion with respect to the occupants of A, and it can cause no changes in the way an event happens in B from the way the event happened in B before this motion with respect to the occupants of B. This is due to the laws of nature in system A with respect to the occupants of A being the same as the laws of nature in system B with respect to the occupants of B even though there is uniform motion between A and B.

The Relativity Principle of Einstein
And the Classical Relativity Principle of Galileo

In *Dialogue on the Great World Systems* of 1632, Galileo describes the classical relativity principle with these words:

Shut yourself up with some friend in the main cabin below decks on some large ship, and have with you there some flies, butterflies, and other small flying animals. Have a large bowl of water with some fish in it; hang up a bottle that empties drop by drop into a wide vessel beneath it. With the ship standing still, observe carefully how the little animals fly with equal speed to all sides of the cabin. The fish swim indifferently in all directions; the drops fall into the vessel beneath; and, in throwing something to your friend, you need throw it no more strongly in one direction than another, the distances being equal; jumping with your feet together, you pass equal spaces in every direction. When you have observed all these things carefully (though there is no doubt that when the ship is standing still everything must happen in this way), have the ship proceed with any speed you like, so long as the motion is uniform and not fluctuating this way and that. You will discover not the least change in all the effects named, nor could you tell from any of them whether the ship was moving or standing still ... The cause of all these correspondences of effects is the fact that the ship's motion is common to all the things contained in it (comments by Salviati on the Second Day).

Hence, like Einstein's relativity principle, Galileo's principle also points out that if system A (a ship) is in uniform motion relative to system B (such as the ocean) after the two systems had been at rest with each other, then this motion can cause no change in the way things happen in A from the way they happened before this motion with respect to the occupants of A.

In chapter 5 of his book, *Relativity: The Special and the General Theory*, Einstein speaks about the exactness of the classical relativity principle "in the domain of mechanics," but he goes on to say that "classical mechanics affords an insufficient foundation for the physical description of all natural phenomena." Einstein then points out in chapters 6 and 7 that in order for the classical relativity principle to be sufficient, it must also take into account electromagnetic waves. Hence, unlike the classical relativity principle, the relativity principle of special relativity insists that when there is uniform motion between system A and system B after the two systems had been at rest with each other, this motion can cause no changes in the way a mechanical or electromagnetic event happens in system A from the way the event happened before this motion with respect to the occupants of A, and it can cause no changes in the way a mechanical or electromagnetic event happens in system B from the way the event happened before this motion with respect to the occupants of B. As will be shown in coming chapters, we will see that the addition of electromagnetic waves to the classical relativity principle enables this principle along with the constant speed of light principle to create time dilation and length contraction.

The Meaning of Uniform Motion
according to Einstein

As shown previously, both Galileo and Einstein claim that the relativity principle functions through "uniform motion." As to what uniform motion means according to Einstein, he gives a definition in chapter 5 of his book *Relativity: The Special and the General Theory*, where, in describing the relativity principle of special relativity, he again speaks about a railway carriage in uniform motion. "We call its motion a uniform translation ('uniform' because of its constant velocity and direction, 'translation' because although the carriage changes its position relative to the embankment yet it does not rotate in so doing)." Hence, according to these words, the uniform motion of a system is motion at a constant velocity and in a constant direction. In chapter 18 of the same book, Einstein points out that uniform motion must also be motion of a system that cannot be felt by the occupants of the system. "If the motion of the carriage is now changed into a non-uniform motion, as for instance by a powerful application of the brakes, then the occupant of the carriage experiences a corresponding powerful jerk forwards." Hence, according to Einstein, if uniform motion changes into motion that can be felt, the uniform motion becomes "non-uniform motion." And as pointed out by Galileo, the relativity principle functions through the motion of a ship "so long as the motion is uniform and not fluctuating this way and that."

The Constant Speed of Light Principle

The second principle underlying special relativity concerns the constant speed of light. In chapter one of "On the Electrodynamics of Moving Bodies," Einstein describes the principle with these words: "Light always propagates in empty space with a definite velocity V that is independent of the state of motion of the emitting body." Hence, according to this principle, every ray of light moves through empty space at the same constant velocity V (now called "c" rather than "V") regardless of any velocity of the ray's source. Such independence from sources makes the speed of light very unique since the velocity of all material objects, such as a ball being thrown forward from a moving train, is determined by adding the speed of the system from which an object is ejected to the speed of the object. Although Einstein's light principle refers specifically to rays of light, it applies to all electromagnetic waves, such as radio waves, X-rays, and gamma waves.

The Practical Application of Einstein's Principles

Although Einstein's two principles in special relativity are clear and specific, it is also apparent that, according to Einstein, the principles can be applied in a practical rather than an absolute sense. To show this, we turn to chapter 4 of "On the Electrodynamics of Moving Bodies." In that chapter Einstein has the principles and earth's rotational motion causing a clock at the equator to run slower than a clock at one of earth's poles.

Although special relativity's constant speed of light principle calls for an empty, vacuous environment, Einstein has this principle functioning in the non-vacuous atmosphere of the earth. This is undoubtedly due to earth's environment having no consequential effect on the speed of light, and hence, this environment can be considered an environment in which the constant speed of light principle functions.

Light Measured at c

From the constant speed of light and the relativity principle of special relativity, the deduction emerges that a beam of unimpeded light moving past an observer has a constant speed of c. Hence, if the speed of a beam could be measured by an observer, the beam must be measured at c regardless of any motion by the observer. Like the principle that all beams of light must move at the same speed despite various speeds of their sources, this deduction also brings to light a very unique feature. In the mechanical world, if car A is moving at a constant speed of 70-miles-per-hour and if car B is moving in the same direction as A at a constant speed of 30-miles-per-hour, then car A will pass car B at 40-miles-per-hour. The speed of the car being passed is subtracted from the speed of the car that is doing the passing. If an occupant of B had a radar gun, he would measure the speed of A relative to B at 40-miles-per-hour even though the speedometer of car A is revealing a speed of 70-miles-per-hour. However, on the basis of a light beam traveling at a constant speed of one-foot-per-nanosecond and a spaceship moving in the same direction as the beam at a speed of one-fourth a-foot-per-nanosecond, the principles of special relativity insist that the beam must pass the spaceship not at a speed of three-fourths a-foot-per-nanosecond but at a speed of one-foot-per-nanosecond. And if the occupants of the spaceship could measure the speed of the beam that was passing the spaceship, they would measure it at one-foot-per-nanosecond. Likewise, if the spaceship were traveling at any other speed or even if the spaceship were moving opposite the direction of the beam, the beam would always pass the spaceship at one-foot-per-nanosecond, and that is the speed at which the occupants would always measure it.

In order to show how the two principles of special relativity create this deduction, we begin by again pointing out that the speed of every unimpeded beam of light is c. And on the basis that c is one-foot-per-nanosecond, this is the speed of every unimpeded beam. Furthermore, by every unimpeded beam having the same speed, if two beams are parallel to each other and moving by each other in the same direction, their speeds must be the same and neither beam can pass by the other beam.

We continue with a spaceship at rest with respect to a space station. Within the spaceship, point A is at the front of the ship and point B is at the back of the ship. The length between points A and B is 100 feet as measured by the spaceship occupants. When a beam of light is fired from point A back to point B and another beam is fired from B forward to point A, synchronized clocks at A and B report to the spaceship occupants an elapsed time of 100 nanoseconds for both beams. Hence, by dividing the 100 foot length over which each of the beams moves by an elapsed time of 100 nanoseconds, the occupants determine that both the A to B and the B to A beams have a speed of one-foot-per-nanosecond or c. Furthermore, if point X is on the ceiling of the spaceship and point Y is diagonally behind point X on the spaceship's floor at a measured length of 80 feet, a beam fired from X to Y and another beam fired from Y to X must have a speed of one-foot-per-nanosecond (as did the beams from A to B and from B to A) because of the constant speed of light. Hence, clocks at X and Y reveal an elapsed time of 80 nanoseconds for the X to Y and the Y to X beams. Thus, the spaceship occupants measure the speed of both beams at one-foot-per-nanosecond by dividing their 80 foot length by an elapsed time of 80 nanoseconds. The same would be the case concerning any other beams fired between fixed points in the spaceship at any angle.

Let us then say that while the spaceship is still at rest with the space station, a beam of light outside the spaceship, beam XX, passes by the spaceship on a pathway that is parallel to the A to B beam and moving in its direction. Since the occupants measured the speed of the A to B beam at one-foot-per-nanosecond, they must also measure beam XX at one-foot-per-nanosecond since beam XX cannot pass by the AB beam at a speed that differs from the AB beam. The same kind of result would occur if a beam outside the spaceship were moving in the direction of any beam in the spaceship and parallel to the beam. The speed of the outside beam would be measured at one-foot-per-nanosecond even as the speed of each beam inside the spaceship is measured at one-foot-per-nanosecond.

We then have the spaceship in uniform motion relative to the space station at a speed of .5-foot-per-nanosecond or .5c. However, despite this enormous speed, the relativity principle of special relativity insists that, with respect to the spaceship occupants, this .5c motion between the spaceship and the space station can cause no changes in the way an event happens in the spaceship from the way the event happened before this motion. Hence, the occupants measure every beam fired in the spaceship at a speed of one-foot-per-nanosecond by measuring the length over which each beam moves and by dividing this length by the elapsed time of the beam as revealed by clocks at both ends of each beam. Then on the basis that beams outside the spaceship move past beams inside the spaceship on pathways that are parallel to and in the direction of the spaceship beams, each of these outside beams would have the same speed as the inside beam, one-foot-per-nanosecond, and would be measured at that speed by the spaceship occupants. Hence, the occupants would end up measuring every unimpeded beam fired within and outside the spaceship at c, one-foot-per-nanosecond, including those moving in the direction of the spaceship, against this direction, and at an angle relative to the direction. This would be the case regardless of any speed or any rest of the spaceship relative to the space station.

Thus, as shown previously, the claims of special relativity through the functioning of its two principles cause the occupants of the spaceship to measure every unimpeded beam of light at c regardless of whether or not the spaceship is at rest or in uniform motion relative to the space station. This must also be the case concerning any other occupants in any other system despite any rest or uniform motion of the system relative to other systems. All of these occupants must measure every unimpeded beam of light at c according to the claims and principles of special relativity.

CHAPTER 2

Preludes to Special Relativity's Length Contraction And Time Dilation

It is with the principles of relativity and light that Einstein proceeds into his 1905 paper "On the Electrodynamics of Moving Bodies" and shows how the two principles can bring about time dilation and length contraction by functioning through uniform motion.

However, before showing how these principles produce these results, we first look at other ideas about the speed of time and the shortening of length that arose before Einstein's 1905 paper and some of the history behind them.

The Michelson-Morley Experiment And the Response of Hendrik Lorentz

As early as the sixteenth century, there were scientists who believed that light traveled in waves. They also thought that it traveled through a substance that they called luminiferous (light-carrying) ether. They reasoned that there can't be waves without something that waves. Since light can move throughout the universe and even through artificially created vacuums, it was assumed that this ether filled the entire universe even though it could not be seen, felt, or weighed. Eventually, however, the particle theory of light gained dominance and the idea of a universal ether faded into the background. But then came the nineteenth century with its amazing discoveries concerning the nature of light and the evidence that light indeed travels in waves. This led, in turn, to a renewed interest in luminiferous ether, the theoretical substance that enabled light waves to exist and travel. And as might be expected, this widespread interest in the ether launched various efforts to find concrete evidence for its existence.

The most famous of these ether-detecting experiments was conducted in 1887 by two American scientists, Albert Michelson and Edward Morley. More information about this experiment will be given in chapter 12 but, for now, we point out that the experiment involved an instrument called an interferometer. It consisted of a five-foot-square block of stone that was floated on a pool of mercury so that it could easily be turned. On the surface of the interferometer was a projector that fired two rays of light. One ray traveled from one corner of the interferometer, say corner A, to the corner farthest from it, C, and the other ray moved from corner B to corner D. Hence, the AC and BD lengths or arms on the interferometer were at a right angle to each other. Mirrors at each corner were used to increase the length pathway of the beams. By the earth and the interferometer passing through the ether at a speed of about 67,100-miles-per-hour as the earth orbited around the sun, the scientists were not only convinced that the ether would cause a slowing of both rays but also that the slowing of the ray on an arm that was parallel to the orbit would be greater than the slowing of the ray on the arm that was at a right angle to this motion. This slowing of both rays would mean an increase in their elapsed times as they moved over these arms and an increase in the elapsed time of the ray on the parallel arm that was greater than the increase of the ray on the right angle arm. Hence, by continuously rotating the interferometer, the AC and BD arms would continuously change from being parallel to being at a right angle to the orbit, and the changing elapsed times of the rays over the two lengths would also be continuous. Michelson and Morley also made calculations as to the amount of change that would occur in these elapsed times by the impact of the ether upon them. However, since the amount of change that actually took place was not even close to what had been calculated, the experiment failed to reveal the existence of the ether.

Michelson and Morley concluded that although the ether existed, their interferometer did not detect it because of the possibility that the ether around the interferometer moved with the instrument. Hence, they suggested that the outcome of their experiment might be different if it were conducted "at the top of an isolated mountain peak." Various other explanations concerning the experiment's result were put forth by other scientists. But possibly the strangest sounding answer came from a Dutch scientist, Hendrik Lorentz, and can be found in "Michelson's Interference Experiment," a paper by Lorentz published in 1895 in Leyden, Holland. A translation of this paper from French into English can be found in *Relativity Theory: Its Origins & Impact on Modern Thought*, a book edited by L. Pearce Williams. Like Michelson and Morley, Lorentz continued to accept the existence of ether, but he added a new twist concerning its effects. His study of atoms led him to the conclusion that "the form and dimensions of a solid body are ultimately conditioned by the intensity of molecular actions." Hence, when a corner of Michelson's stone block was pointed in the direction of the earth's movement around the sun, its push against the ether caused a molecular compression and this compression caused a very slight decrease in the stone's length from what it would be without this motion. This compressed length would be parallel to the interferometer's orbital motion and extend from the corner of the interferometer moving into the ether to the opposite corner. However, by the length between the other two corners being at a right angle to this motion, it would undergo no shortening. Lorenz agreed with Michelson and Morley that, if this compression of the stone did not occur, then there would be a greater increase in the elapsed time of a beam over a length parallel to the stone's motion than over a length at a right angle to the motion. However, by there being a physical shortening of the parallel length, the increased elapsed times of both beams could be the same. Hence, Lorentz wrote, "If we assume the arm which lies in the direction of the earth's motion to be shorter than the other, then the result of the Michelson experiment is explained completely."

Thus, prior to the introduction of special relativity in 1905, it was being theorized that the motion of a system through space caused a physical contraction in the length of the system. When special relativity arrived, it also claimed that the motion of a system through space would cause a contraction in the length of the system. However, the cause of such a contracted length as claimed by Lorentz differed from the cause as claimed by special relativity.

The Lorentz-Fitzgerald Contraction, $\sqrt{1-v^2/c^2}$

The theory that Lorentz presented to explain the failure of the Michelson-Morley experiment not only included the claim concerning a contraction in the length of the interferometer that fronted its motion through the ether, but the theory of Lorentz also presented a formula that could be used to determine specific amounts for changes that were caused by the ether and the velocity of the interferometer. It reads $\sqrt{1-v^2/c^2}$, the square root of one minus the velocity of a system squared divided by the constant speed of light squared. As shown by Lorentz, if the amount by which the length of the interferometer decreased were its at-rest length multiplied by $\sqrt{1-v^2/c^2}$, then the amount revealed by this multiplication would be the exact amount that was needed to explain the results of the Michelson-Morley experiment.

Six years before 1895, the year in which Lorentz published the "Michelson Interference Experiment" with its length contraction theory and the $\sqrt{1-v^2/c^2}$ formula, a short letter to the editor of a newspaper by an Irish physicist named George Fitzgerald, titled "The Ether and the Earth's Atmosphere," appeared in *Science* magazine (1889). In this letter Fitzgerald offers an explanation for the results of the Michelson-Morley experiment that corresponded to the one given by Lorentz. He wrote that "the length of material bodies changes, according as they are moving through the ether or across it, by an amount depending on the square of the ratio of their velocity to that of light." He later expressed similar ideas in the July 19, 1900, issue of *Nature* magazine in an article titled "The Relations Between Ether and Matter." This article now appears in *The Scientific Writings of the Late George Francis*

Fitzgerald, edited in 1902 by Joseph Larmor as number 97. Shortly after the1889 appearance of the Fitzgerald letter in the *Science* magazine, Lorentz sent a brief note to Fitzgerald informing him that two years earlier he had arrived at the same conclusion and that his conclusion had been published in the Proceedings of the Dutch Academy of Sciences. Enclosed with his note were copies of the relevant documents. In response, Fitzgerald sent a letter to Lorentz in which he spoke about a much earlier letter that he had sent to *Science* about his length contraction theory, but he did not know if it ever got published. Hence, these letters indicate that the two scientists arrived at conclusions that were independent of each other concerning the use of $\sqrt{1-v^2/c^2}$ in explaining the results of the Michelson-Morley experiment. Copies of these letters can be found in an article by Stephen G. Brush, "Note on the History of the Fitzgerald-Lorentz Contraction," which appears in *Isis* (Summer 1967). Today, the $\sqrt{1-v^2/c^2}$ formula is referred to as the Lorentz-Fitzgerald Contraction.

Thus, the $\sqrt{1-v^2/c^2}$ formula was originally used to determine the amount by which the ether caused a physical contraction in the length of the parallel arm on the Michelson-Morley interferometer. When special relativity arrived a few years later, $\sqrt{1-v^2/c^2}$ was used to determine the amount by which the motion of system A relative to observers in system B caused the elapsed time of a returned beam in system A over a vertical length to increase, to determine the amount by which time in system A was passing slower than in system B, and to determine the amount by which the length of system A is contracted with respect to system B.

The "Local Time" of Hendrik Lorentz and Henri Poincaré

In addition to his "Michelson Interference Experiment" of 1895, Lorentz published another paper that same year, "Inquiry into a Theory of Electrical and Optical Phenomena in Moving Bodies." A translation of this paper from French into English by wikisource can be found on the internet. In this paper Lorentz spoke about "local time." He connected this concept to his observation that the amount of time required for light to reveal an event occurring in system X to the occupants of system Y cannot be determined only on the basis of the distance between X and Y when the event occurs. It must also be determined on the basis of any motion of X that causes a change in the distance that the light must cover as it travels from X to Y.

Although scientists reacted in various ways to the writings of Lorentz, there were certainly those who enthusiastically embraced his contraction of length and local time concepts. Among them was the French mathematician and theoretical physicist Henri Poincaré as shown by his paper "The Principles of Mathematical Physics" of 1904. A French-to-English translation of this article can be found in *Relativity Theory: Its Origins & Impact on Modern Thought*, a book edited by L. Pearce Williams. In this paper Poincaré wrote:

> All attempts to measure the velocity of the earth in relation to the ether have led to negative results ... The means have been varied in a thousand ways and finally Michelson has pushed precision to its last limit; nothing

has come of it. It is precisely to explain this obstinacy that the mathematicians are forced today to employ all their ingenuity. Their task was not easy, and if Lorentz has gotten through it, it is only by accumulating hypotheses. The most ingenious idea has been that of local time.

In order to illustrate this ingenious idea, Poincaré wrote about two clocks, clock A at station A and clock B at station B. In order to determine if the clocks are synchronized, station A sends a signal to station B when clock A displays an agreed-upon hour, such as 3:00 p.m., and station B determines when the signal arrives according to clock B. We'll say that, in this case, the signal came to B at 10 microseconds after 3:00 p.m. Station B then sends a signal to station A at a designated hour, such as 4:00 p.m., and if clock A reveals that the signal arrived at 10 microseconds after 4:00, then the clocks are synchronized. And by the two clocks being synchronized and revealing the same elapsed time for the signals between them, the clocks are marking "the true time."

However, according to Poincaré, results would change if the system on which the two stations and clocks were located went into motion along a line that was parallel to the line be-tween the two clocks. In this case, by a signal being sent from station A to station B in the direction of the system's move-ment, B would move away from A as the signal traveled to B, and this would increase the signal's elapsed time. We'll say that instead of clock B revealing a passing of 10 microseconds for the signal's journey, clock B is now reporting an elapsed time of 12 microseconds for the journey. On the other hand, by a signal being sent from station B to station A against the direction of the system's movement, A would move toward B as the signal traveled to A, and this would reduce the signal's elapsed time from 10 to about 8 microseconds as reported by clock A. Poincaré then writes that clocks "adjusted in that manner do not mark, therefore, the true time; they mark what one may call the *local time*, so that one of them goes slow on the other." Thus, according to Lorentz and Poincaré, motion can be involved in time being "true time" or in time being "local time."

Poincaré then follows his presentation on true time and local time with what he calls a complementary hypothesis. "It is necessary to admit that bodies in motion undergo a uniform contraction in the sense of the motion. One of the diameters of the earth, for example, is shrunk by $1/200,000,000^{th}$ in consequence of the motion of our planet, while the other diameter retains its normal length." Hence, in the year prior to Einstein's "On the Electrodynamics of Moving Bodies," Poincaré was writing about how motion can cause the contraction of an object's length and cause times that differ with one another.

CHAPTER 3

Special Relativity's Time Dilation and Length Contraction

As shown in the last chapter, theories were being developed during the years just prior to 1905 concerning the way in which motion can bring about a "local time" that differed from "true time" and also bring about a contraction of a system's length. Then in 1905, Einstein published his paper "On the Electrodynamics of Moving Bodies" and introduced his theory of special relativity. In this paper Einstein also spoke about how motion can bring about times that differed with one another and cause the contraction of systems. However, these time and length phenomena resulted from the principles of special relativity functioning through uniform motion and differed significantly from all previous ideas and claims. In this chapter, we will show how the claims and principles of special relativity can bring about these two phenomena. And as pointed out in the introduction to this book, we will be using one-foot-per-nanosecond as the constant speed of light.

Time Dilation

Concerning the claim of time dilation by special relativity, we will conduct an experiment that begins with uniform motion between spaceship X and spaceship Y at a speed of .5c. The length of spaceship X as measured by its occupants is 100 feet and, within spaceship X, there is a 20 foot length between point A on the floor of the spaceship and point B on the spaceship ceiling that is vertically above point A. A pulse of light is then fired in spaceship X from a source at point A that travels over the 20 foot length from point A to point B and is reflected from a mirror at point B back to point A.

In accord with the relativity principle of special relativity, none of the .5c uniform motion between spaceship X and spaceship Y is the motion of spaceship X with respect to the occupants of spaceship X. Hence, the pulse of light from point A travels 20 feet from point A to point B and 20 feet from B to A for a total distance of 40 feet and an elapsed time of 40 nanoseconds relative to the spaceship X occupants.

Also in accord with the relativity principle, all of the .5c uniform motion between spaceship X and spaceship Y is the motion of spaceship X with respect to observers in spaceship Y. This motion of spaceship X relative to the spaceship Y observers causes point B on the ceiling to move away from the pulse that is fired from point A. Hence, when the pulse reaches point B, it is on a diagonal pathway because of the constant speed of light principle not allowing the .5c speed of the source at point A to be added to the pulse. Then after traveling from A to B on a diagonal pathway, the pulse travels on a diagonal pathway from B to A, which has the same length as the A to B diagonal pathway. Hence, the pulse forms the sides of an isosceles triangle relative to the observers in spaceship Y. Concerning the elapsed time of this triangular pulse, a determination can be made by dividing the 40 foot A to B to A length over which the pulse moves by the Lorentz-Fitzgerald contraction, $\sqrt{1-v^2/c^2}$. Thus, by the v in this contraction being the .5c velocity of the spaceship and by c being the 1.0c speed of light, the contraction reads .866. Then by dividing the 40 foot ABA length by .866, the pulse is shown to have an elapsed time of 45.19 nanoseconds with respect to the spaceship Y observers.

Thus, during the .5c uniform motion between spaceship X and spaceship Y, the pulse in spaceship X has an elapsed time of 40 nanoseconds with respect to the occupants of spaceship X. However, with respect to the observers in spaceship Y, the .5c motion of spaceship X has caused this same pulse to have an elapsed time of 45.19 nanoseconds. Thus, with respect to the spaceship Y observers, the pulse in spaceship X is revealing that the motion of spaceship X is causing the rate of time in spaceship X to be slower than in spaceship Y and slower than what it would be without this motion by factor of .866 (40 ÷ 45.19 = .866). When 45.19 nanoseconds pass in spaceship X with respect to the observers in spaceship Y, only 40 nanoseconds pass in spaceship X with respect to the occupants of spaceship X.

Length Contraction

Having shown how the claims of special relativity and the .5c motion of spaceship X cause time in spaceship X to dilate by a factor of .866 with respect to the observers in spaceship Y, our next task is to show how the claims of special relativity and the .5*c* motion of spaceship X with respect to the space-ship Y observers cause a corresponding .866 contraction in the length of spaceship X. To do this, we begin by pointing out that spaceship X has a length of 100 feet with respect to the occupants, an amount determined by this length being measured by the occupants. As to the length of spaceship X with respect to the observers in spaceship Y, this length can be determined by having spaceship X fly past a marker on spaceship Y. Then, on the basis that there is no such thing as time dilation and length contraction, it would take 200 nanoseconds for the entire length of spaceship X to fly past this marker by moving at a speed of .5*c*. Hence, with respect to the spaceship Y observers, spaceship X would have a length of a hundred feet (*200 nanosecond elapsed time x .5c speed = 100 foot length*). However, by time in spaceship X dilating by a factor of .866 with respect to the observers in spaceship Y and in accord with the claims and principles of special relativity, the elapsed time for spaceship X to move past the marker is not two hundred nanoseconds, but 173.2 nanoseconds (200 x .866 = 173.2). Hence, the previous equation, *200 nanosec-ond elapsed time x .5c speed = 100 foot length*, now reads *173.2 nanosecond elapsed time x .5c speed = 86.6 foot length*. Thus, by the .5*c* speed of spaceship X and the claims of special rela-tivity causing time in spaceship X to dilate by a factor of .866 with respect to the spaceship Y observers, this speed and the claims of special relativity also cause the 100 foot measured length of spaceship X to be contraction by a factor of .866 (86.6 ÷ 100 = .866) with respect to the spaceship Y observers.

Let us also say that, with respect to the spaceship Y ob-servers, spaceship X is traveling over a fixed length from point A to B at a speed of .5*c*. Let us also say that the distance from point A to B as measured by the occupants of spaceship X is 1,000,000 feet. Hence, with respect to the spaceship Y observers, it would take 2,000,000 nanoseconds for spaceship X to pass over this length if there were no such thing as time dilation and length contraction (*1,000,000 foot AB length ÷ .5c speed = 2,000,000 nanoseconds*). However, by time in spaceship X dilating by a factor of .866 with respect to the spaceship Y observers, the elapsed time for the spaceship to move from A to B is not two million nanoseconds, but 1,732,000 nanoseconds (*2,000,000 nanoseconds x .866 = 1,732,000 nanoseconds*). Hence, we now have this equation: *1,732,000 nanosecond elapsed time x .5c speed = 866,000 foot AB length.* Thus, by the .5c speed of spaceship X and the claims of special relativity causing time in spaceship X to dilate by a factor of .866, this .5c speed and the claims of special relativity cause a corresponding .866 contraction of a fixed length over which spaceship X moves relative to the spaceship Y observers (866,000 feet ÷ 1,000,000 feet = .866).

However, the claims of special relativity and the .5c motion between spaceship X and spaceship Y cause no dilation of time in spaceship X or a contraction in the length of spaceship X relative to the occupants of spaceship X because, in accord with the relativity principle, none of this .5c motion is the motion of spaceship X with respect to occupants of spaceship X. Hence, there is no motion of spaceship X to cause these results.

The Reciprocal Results of Special Relativity's Principles

The way in which the claims and principles of special relativity brought about time dilation in spaceship X and a contraction in the length of spaceship X relative to the observers in spaceship Y was based on the .5c uniform motion between these two system being the motion of spaceship X with respect to the observers in spaceship Y. However, in accord with the relativity principle, the .5c uniform motion between spaceship X and spaceship Y is also the motion of spaceship Y relative to the observers in spaceship X. Hence, even as this .5c motion and the principles of special relativity caused a dilation of time in spaceship X and a contraction in the length spaceship X by a factor of .866 relative to the spaceship Y observers, so also does this motion and the principles cause time dilation in spaceship Y and a contraction in the length of spaceship Y by a factor of .866 relative to the spaceship X observers.

Chapter 4

Evidence Refuting the Claims of Special Relativity From Clock Synchronization Approaches

In this chapter, two clock synchronization approaches will be presented - the Einstein Synchronization Approach and the Revised Synchronization Approach. It will be shown how both approaches take place in accord with the claims and principles of special relativity. However, it will also be shown how results produced by these two approaches are in conflict with each other. Hence, by these conflicting results being caused by the claims and principles of special relativity, these results will bring evidence against these claims and principles.

As stated in the introduction to this book, we will be using one-foot-per-nanosecond as the constant speed of light even though the actual speed is .983571-foot-per-nanosecond.

The Einstein Synchronization Approach

Material concerning the Einstein Synchronization Approach comes from chapters 1 and 2 of Einstein's "On the Electrodynamics of Moving Bodies" article. In chapter 1, Einstein shows how clock A and clock B can be determined synchronous with respect to observers at rest with the clocks by a ray of light being fired from clock A to clock B and reflected from clock B to clock A. According to Einstein, "The clocks are synchronous by definition if $tB - tA = t'A - tB$ (*the time of day on clock B minus the time of day on clock A equals the time of day on clock A minus the time of day on clock B*)." Einstein then adds, "We assume that it is possible for this definition of synchronism to be free of contradiction."

However, before declaring that clocks A and B must be synchronized if $tB - tA = t'A - tB$, we point out that there is only one time of day on clock B in the clock synchronization equation, $tB - tA = t'A - tB$. This one time of day on clock B is when the ray arrives at clock B from clock A and leaves clock B in returning to clock A. Although very unlikely, it could happen that clock B is not synchronized with clock A and that it is only by chance that, when the ray arrives at clock B and leaves clock B, the equation reads $tB - tA = t'A - tB$. As an example, let us say that clocks A and B are not synchronized. However, when the synchronization of clocks A and B is tested, the equation reads tB *(7:50 a.m.)* - tA *(7:00 a.m.)* = $t'A$ *(7:100 a.m.)* - tB *(7:50 a.m.)*. (In this chapter, a time of day such as 7:100 a.m. means 100 nanoseconds after 7:00 a.m.) Hence, although the time of day on clock B is 7:50 a.m. when the ray comes to clock B and leaves clock B, the reason is not that clock B is synchronized with clock A. The reason is that, as a matter of pure chance, clock B happens to be posting 7:50 a.m. when the ray arrives and leaves clock B. Although it is very unlikely that such a chance occurrence would happen, it could happen. And if it did happen, the

Einstein Synchronization Approach would erroneously be claiming that clocks A and B are synchronized. Hence, if clocks are needed that are synchronized with certainty, then clocks that are shown to be synchronized by the Einstein Synchronization Approach cannot be used.

The Einstein Synchronization Approach as presented in chapters one and two of the electrodynamics article continues with clocks A and B at points A and B at the two ends of a rod and with the rod being in uniform motion with respect to commoving observers. During this motion, a ray is fired from a source at point A that travels from clock A forward to clock B and is reflected from clock B to clock A. It is then pointed out that the motion of the rod causes an increase in the elapsed time of the A to B ray with respect to the commoving observers in accord with the constant speed of light principle. The amount of this increased elapsed time can be determined by the use of this equation, $tB - tA = rAB \div c - v$ (the elapsed time of the ray from clock A to clock B = the AB length of the rod divided by the speed of light minus the velocity of the rod and the clocks on the rod). Concerning the journey of the ray from clock B back to clock A, it is pointed out that the motion of the rod causes a decrease in the elapsed time of the B to A ray with respect to the commoving observers in accord with the constant speed of light principle. The amount of this decreased elapsed time can be determined by the use of this equation, $tA - tB = rAB \div c + v$ (the elapsed time of the ray from clock B to clock A = the AB length of the rod divided by the speed of light plus the velocity of the rod and the clocks on the rod).

Thus, with respect to the commoving observers, the elapsed time of the ray from clock A to clock B is greater than the elapsed time of the ray from clock B to clock A. In other words, tB - tA does not equal t'A - tB. Hence, by tB - tA not being the same as t'A - tB, this means that, according to the clock synchronization equation, $tB - tA = t'A - tB$, the clocks are not synchronized relative to the commoving observers.

Hence, it is claimed by the Einstein Synchronization Approach that clocks A and B on the two ends of the rod cannot be synchronized with respect to observers before whom the rod and the clocks are in uniform motion. It is also claimed by the Einstein Synchronization Approach that clocks A and B are synchronized with respect to observers before whom the rod and the clocks are at rest.

Einstein then closes chapter 2 with the following observations:

> Thus we see that we cannot ascribe absolute meaning to the concept of simultaneity; instead, two events that are simultaneous when observed from some particular coordinate system can no longer be considered simultaneous when observed from a system that is moving relative to the system.

However, before ending this section concerning the Einstein Synchronization Approach, we again point out that clocks A and B are not synchronized with certainty if $tB - tA = t'A - tB$. Hence, anything revealed by this equation concerning clock synchronization or the concept of simultaneity is not certain.

The Revised Synchronization Approach

The Revised Synchronization Approach takes place in accord with the claims of special relativity and begins with spaceship X and spaceship Y at rest with each other. Within spaceship X, clock A is in the back of the spaceship and clock B is in the front of the spaceship. Furthermore, two rays of light will be fired in the Revised Synchronization Approach, ray one and ray two. Ray one will travel from clock A to clock B to clock A, and ray two will travel from clock B to clock A to clock B. In what follows, we will show how the firing of these two rays can make it certain that clocks A and B are synchronized.

We begin with clocks A and B in spaceship X not being synchronized in that the time of day on clock B is behind the time of day on clock A even though clock B is running faster than clock A. Hence, eventually the moment happens in which the time of day on clock B catches up with the time of day on clock A. Thus, at that moment, the time of day on clock A is the same as the time of day on clock B. Furthermore, by ray one being fired from clock A that reaches clock B when the times of day on the two clocks are the same, the elapsed time of ray one from clock A to B is the same as the elapsed time of ray one from clock B to clock A. Hence, at that moment $tB - tA$ equals $t'A - tB$. Thus, the equation seems to be revealing that the clocks are synchronized even though this is not the case. We will call this moment the artificial synchronization moment. It then happens after the artificial synchronization moment that the time of day on clock B is ahead of the time of day on clock A and remains ahead of clock A because clock B is running faster that clock A.

Several seconds after the artificial synchronization moment, ray two is fired in spaceship X from clock B to clock A to clock B. Hence, by clock B running faster than clock A, the clocks must reveal through the firing of ray two that the elapsed time of ray two from clock B to clock A is not the same as the elapsed time of ray two from clock B to clock A and that *tA - tB does not equal t'B - tA*. Hence, ray two makes it certain that clocks A and B are not synchronized and that it was an artificial synchronization moment that caused tB - tA to equal t'A - tB during the firing of ray one.

However, let us say that when ray one was fired, clocks A and B were synchronized. Hence, this was the reason that *tB - tA equaled t'A –tB* when ray one made its journey from A to B to A. Thus, when ray two is fired a few seconds after ray one, the journey of ray two from clock B to clock A to clock B must reveal that *tA - tB equals t'B - tA*. By clocks A and B being synchronized when ray one was fired, they must also be synchronized when ray two is fired.

Thus, as shown above, if the A to B to A journey of ray one causes *tB - tA to equal t'A-tB* and if the B to A to B journey of ray two causes *tA - tB to not equal t'B - tA,* then it is certain that clocks A and B are not synchronized. On the other hand, if the A to B to A journey of ray one causes *tB - tA to equal t'A - tB* and if the B to A to B journey of ray two causes *tA - tB to equal t'B - tA,* then it is certain that clocks A and B are synchronized.

At this point, we continue our presentation of the Revised Synchronization Approach with it being shown that clocks A and B are synchronized with certainty. This means that the times of day on clocks A and B reveal that the elapsed time of ray one from clock A to clock B is the same as the elapsed time of ray one from clock B to clock A and that $tB - tA$ *equals* $t'A - tB$. In like manner, it means that the times of day on clocks A and B reveal that the elapsed time of ray two from clock B to clock A is the same as the elapsed time of ray two from clock A to clock B and that $tA - tB$ *equals* $t'B - tA$. Furthermore, each of these equations is revealing that these clocks are fully synchronized in that the times of day on clock A are continuously the same as the times of day on clock B and that the rate of clock A is the same as the rate of clock B.

In order to show how the $tB - tA = t'A - tB$ equation reveals that the times of day on the two clocks are continuously the same, we look at the $tB - tA = t'A - tB$ equation with specific times of day attached to it. Hence, we will have 6:60 on clock B - 6:00 on clock A = 6:120 on clock A - 6:60 on clock B (as previously pointed out, a time of day such as 6:60 means 60 nanoseconds after 6:00.) Thus, according to this equation, when the time of day on clock B is 6:60, the time of day on clock A must be 6:60. As to why this must be the case, let us say that when the time of day on clock A is 6:60, the time of day on clock B is 6:50. Hence, the previous equation would read as follows: 6:50 on clock B - 6:00 on clock A = 6:120 on clock A - 6:50 on clock B or 50 = 70. Of course, this equation is invalid since 50 does not equal 70. However, this invalid equation does reveal that in order for $tB - tA$ to equal $t'A - tB$, the times of day on clock A must continually be the same as the times of day on clock B.

Furthermore, by the clock synchronization equation reading $tB - tA = t'A - tB$, this means that the elapsed time of a ray traveling from clock A to clock B equals the elapsed time of the ray traveling from clock B to clock A. And with this being the case, the equation is revealing that the rate of clock A is the same as the rate of clock B. If the rate of clock B were greater than the rate of clock A, then the elapsed time of the ray from clock A to clock B would be greater than the elapsed time of the ray from clock B to clock A. And if the rate of clock B were slower than the rate of clock A, then the elapsed time of the ray from clock A to clock B would be less than the elapsed time of the ray from clock B to clock A. Hence, by the equation reading $tB - tA = t'A - tB$, this means that the rate of clock A must be the same as the rate of clock B.

We also point out that all of the events and results that are described previously take place while spaceship X and spaceship X are at rest with each other. Hence, all of these events are occurring with respect to the occupants in spaceship X and the observers in spaceship Y. We will also say that during this rest period, the A to B length of spaceship X as measured by its occupants is 60 feet with respect to personnel in spaceship X and spaceship Y. Hence, when a ray is fired in spaceship X from A to B to A or from B to A to B, the ray has an elapsed time of 120 nanoseconds with respect to personnel in both spaceships.

Events in Spaceship X During the Uniform Motion between Spaceships X and Y

Following the rest period between spaceships X and Y, there is uniform motion between them at a speed of .5c. With respect to the occupants of spaceship X and in accord with the principles of special relativity, none of this motion is the motion of spaceship X. Hence, when ray one and ray two are fired in spaceship X over the 60 foot length between clocks A and B, it is revealed that the clocks are fully synchronized in that they are running at the same rates and are continuously posting the same times of day even as this was the case before this .5c motion.

However, with respect to the observers in spaceship Y, the .5c motion between spaceships X and Y is completely the motion of spaceship X. Hence, in accord with the principles of special relativity, this .5c speed of spaceship X causes a contraction of the 60 foot A to B length of the spaceship by a certain factor and also causes a slowing of time in spaceship X by that same factor. This common factor can be determined by the use of the Lorentz-Fitzgerald contraction, $\sqrt{1-v^2/c^2}$. Hence, by v being .5c and by c being 1.0, the contraction reads .866. Then by multiplying the 60 foot, at-rest, A to B length of spaceship X by .866, we have a contracted length of 52 feet (51.96 feet rounded off) with respect to the spaceship Y observers.

We then have ray one being fired in spaceship X from clock A at point A. Hence, by spaceship X having a speed of .5c with respect to the spaceship Y observers, this speed of spaceship X causes clock B at point B to move away from ray one as the ray is traveling toward clock B from clock A because of the constant speed of light principle not allowing the speed of the spaceship and the source of the ray at point A to be added to the ray. This motion of clock B away from ray one causes an increase in the elapsed time of the ray in order for it to reach clock B. The amount of this increased elapsed time can be determined by the equation, $tB - tA = AB \div c - v$ (the elapsed time of ray one from clock A to clock B equals the 52 foot AB length of spaceship X divided by the speed of light minus the velocity of the spaceship). Thus, by the velocity of spaceship X being .5c and the speed of light being 1.0c, the equation reveals that ray one has an elapsed time of 104 nanoseconds in order for it to travel from A to B (1.0c - .5c = .5 and 52 feet ÷ .5 = 104 nanoseconds). Concerning the journey of ray one from clock B back to clock A, the .5c speed of spaceship X causes clock A to move toward the ray as the ray is traveling toward clock A from clock B because of the constant speed of light principle not allowing the speed of the spaceship and the mirror source of the ray to be subtracted from the ray. This motion of clock A toward the ray causes a decrease in the elapsed time of the ray in its journey from clock B back to clock A. The amount of this decrease can be determined by the equation $tA - tB = AB ÷ c + v$ (the elapsed time of ray one from clock B to clock A equals the 52 foot AB length of the spaceship divided by the speed of light plus the velocity of the spaceship). Thus, by the velocity of spaceship X being .5c and the speed of light being 1.0c, the equation reveals that ray one has an elapsed time of 34.67 nanoseconds in order for it to travel from B to A (1.0c + .5c = 1.5 and 52 feet ÷ 1.5 = 34.67 nanoseconds). Hence, when ray one is fired in spaceship X from point A over the 104 foot ABA length and returns to point A, it has an elapsed time of 138.67 nanoseconds (104 + 34.67 = 138.67) with respect to the observers in spaceship Y.

Furthermore, since the elapsed time of ray one in its A to B to A journey in spaceship X was 120 nanoseconds with respect to the spaceship Y observers when spaceship X was at rest with spaceship Y, this means that the motion of spaceship X has caused time in spaceship X to dilate by a factor of .866 with respect to the observers in spaceship Y (120 ÷ 138.67 = .866). Thus, even as the motion of spaceship X has caused the length of spaceship X to contract by a factor of .866 with respect to the spaceship Y observers, this motion of spaceship X has also caused rate of time in spaceship X to be slowed by a factor of .866 from what that rate was before this motion. And by the rate of time in spaceship X being slowed by a factor of .866, the rate of clocks A and B in spaceship X are also slowed by a factor of .866. Moreover, since the rate of clock A was the same as the rate of clock B when the clocks were at rest with respect to the spaceship Y observers, the clocks continue to run at the same rates with respect to the spaceship Y observers because the rates of both clock were slowed by the same .866 factor.

Shortly after the firing of ray one from clock A at point A, ray two is fired from clock B at point B in spaceship X. With respect to observers in spaceship Y, ray two travels from clock B over the 52 foot contracted length from clock B back to clock A and is returned from clock A to clock B. Hence, with respect to the observers in spaceship Y, the .5c speed of spaceship X and the 1.0c speed of light cause the elapsed time of ray two in its journey from clock B to clock A and from clock A forward to clock B to be the same as the elapsed time of ray one in its journey from clock A forward to clock B and back to clock A. Thus, even as ray one had an elapsed time of 138.67 nanoseconds in its journey from clock A to clock B to clock A with respect to the spaceship Y observers, so also does ray two have an elapsed time of 138.67 nanoseconds in its journey from clock B to clock A to clock B with respect to the spaceship Y observers. Hence, the result of firing ray two is the same as the result in firing ray one. With respect to the spaceship Y observers, ray two reveals that the rate of clocks A and B in the spaceship are slowed by a factor of .866. Thus, since the rate of clock A was the same as the rate of clock B before the .5*c* motion between the clocks with respect to the spaceship Y observers, the rate of clock A remains the same as the rate of clock B since both clocks were slowed by the same .866 factor.

Having considered the rates of clocks A and B during the .5*c* motion between the clocks, we next consider the times of day on clocks A and B in spaceship X with respect to observers in spaceship Y. However, we begin by calling attention to the fact that, with respect to the occupants of spaceship X, the times of day on clock A are continuously the same as the times of day on clock B even as this was the case before the motion between spaceships X and Y. We then point out that the times of day on clocks A and B in spaceship X with respect to the occupants of spaceship X must also be the times of day on clocks A and B in spaceship X with respect to the observers in spaceship Y. To demonstrate why this must be the case, let us say that the time of day on clock A is 7:16 a.m. with respect to the occupants of spaceship X. However, the time of day on clock A cannot be 7:16 a.m. with respect to the occupants of spaceship X while being something else such as 9:34 a.m. with respect to the observers in spaceship Y. If the time of day on clock A is 7:16 a.m. with respect to the occupants of spaceship X, then the time of day on clock A is 7:16 a.m. with respect to the observers in spaceship Y. Thus, even as the time of day on clock A is the same as the time of day on clock B with respect to the occupants of spaceship X, the times of day on the two clocks are also the same with respect to the observers in spaceship Y.

Thus, as shown in the two previous paragraph, clocks A and B in spaceship X are fully synchronized not only with respect to the occupant of spaceship X but also with respect to the observers in spaceship Y in that the two clocks are running at the same rate and are continuously posting the same times of day.

However, for the sake of clarification, we ask this question: How can the times of day on clocks A and B in spaceship X continuously be the same with respect to the occupants of spaceship X and the observers in spaceship Y during the .5c motion between the spaceships while this .5c motion is causing the rates of clocks A and B with respect to the spaceship Y observers to be slower than with respect to the spaceship X occupants?

In responding to this question, we begin with the times of day on clocks A and B in spaceship X continuously being the same with respect to the occupants of spaceship X and the observers in spaceship Y during the .5c motion between the two spaceships.

Let us then say that, with respect to the occupants in spaceship X and the observers in spaceship Y, the time of day on clock A in spaceship X is 8:00 a.m. when ray one leaves clock A. Then by the at rest A to B length of spaceship X being 60 feet, when ray one completes its 120 nanosecond journey from A to B to A, the time of day on clock A is 8:120 a.m.(120 nanoseconds after 8:00 a.m.) relative to the personnel in both spaceships. However, it was previously shown that, with respect to the spaceship Y observers, the .5c motion between the two spaceships causes ray one in spaceship X to have an elapsed time of 138.67 in traveling from A to B to A. Hence, by the times of day on clock A revealing an elapsed time of 120 nanoseconds for ray one that has an elapsed time of 138.67 nanoseconds, this means that, with respect to the spaceship Y observers, the .5c motion between the space-ships is causing the rate of clock A to be slower than before this motion and slower than with respect to the spaceship X occupants by a factor of .866 (120 ÷ 138.67 = .866). And what is said about clock A in this paragraph can also be said about clock B.

Conflict between the Einstein Synchronization Approach And the Revised Synchronization Approach

In this chapter, two clock synchronization approaches were presented, the Einstein Synchronization Approach and the Revised Synchronization Approach. Both approaches took place in accord with the claims and principles of special rela-tivity. Concerning the Einstein Synchronization Approach, it reveals through the firing of a single ray that two clocks in a system can very likely be synchronized relative to observers at rest with the clocks, but they very likely cannot be synchro-nized with respect to observers before whom the clocks are in uniform motion. Concerning the Revised Synchronization Approach, it reveals through the firing of two rays that two clocks in a system can, with certainty, be synchronized with respect to observers at rest with the clocks as well as with respect to observers before whom the clocks are in uniform motion. Hence, by the claims and principles of special relativ-ity revealing through the Einstein Synchronization Approach the likelihood that clocks cannot be fully synchronized with respect to observers before whom the clocks are in uniform motion and revealing through the Revised Synchronization Approach that clocks can be fully synchronized with respect to observers before whom the clocks are in uniform motion, we see the claims and principles of special relativity producing results that are in conflict with each other. And by creating results that conflict with each other, the claims and principles of special relativity are invalidating themselves.

Added Note

When a special relativity experiment takes place, it might be helpful to have clocks in the experiment that are synchronized, not only with respect to observers before whom the clocks are at rest but also with respect to observers before whom the clocks are in uniform motion. Hence, as shown by the Revised Synchronization Approach, such clocks are available.

CHAPTER 5

Evidence Refuting the Claims of Special Relativity From a Vertical and Diagonal Pulse of Light

The Role of Clocks in Special Relativity

In order to consider the role of clocks in special relativity, we have uniform motion between spaceship X and spaceship Y at a speed of .5c. Within spaceship X, point A is on the floor of the spaceship and point AA is vertically above point A at a distance of 16 feet as measured by the occupants of spaceship X. Hence, when a returned pulse of light is fired from point A, it travels 32 feet from A to AA to A and has an elapsed time of 32 nanoseconds with respect to the occupants of spaceship X on the basis that the speed of light is one-foot-per-nanosecond. Let us then say that there is a clock at point A in spaceship X that is visibly revealing to the spaceship X occupants an elapsed time of 32 nanoseconds for the A to AA to A pulse.

With respect to the observers in spaceship Y, the .5c motion between spaceship X and spaceship Y is entirely the motion of spaceship X. This motion of spaceship X causes point

AA to move away from the pulse as it travels from the projector at A to AA due to the constant speed of light principle not allowing the speed of the spaceship and the projector to be added to the pulse. This motion of AA away from point A causes the distance and elapsed time of the pulse to increase over what they would be without this motion and, in returning to point A, there is the same increase in the distance and elapsed time of the pulse. Thus, the pathway of the pulse forms the sides of an isosceles triangle. The elapsed time of the isosceles pulse can then be determined by dividing the 32 foot vertical A to AA to A length over which the pulse moves by $\sqrt{1\text{-}v^2/c^2}$, the Lorentz-Fitzgerald contraction. Hence, with v being .5c and c being 1.0c, the contraction reads .866. When the 32 foot length is divided by .866, the pulse is shown to have an elapsed time of 36.95 nanoseconds relative to the spaceship Y observers.

Thus, by the pulse in spaceship X having an elapsed time of 32 nanoseconds with respect to the occupants of spaceship X and by this same pulse have an elapsed time of 36.95 nanoseconds with respect to the observers in spaceship Y, the pulse is revealing that the rate time in spaceship X is slower than the rate of time in spaceship Y by a factor of .866 (32 ÷ 36.95 = .866) relative to the spaceship Y observers.

Concerning the length of spaceship X with respect to the spaceship Y observers, the dilation of time in spaceship X by a factor of .866 causes the length of spaceship X to contract by that same .866 factor. Hence, the 100 foot length of spaceship X as measured by its occupants is contracted to a length of 86.6 feet with respect to the spaceship Y observers (100 x .866 = 86.6). Let us then say that there is a clock in spaceship Y that is giving information to the spaceship Y observers concerning the time dilation and the length contraction in spaceship X.

Following what is revealed by the clock in spaceship X and the clock in spaceship Y during the .5c motion between the two systems, let us say that the two clocks are tossed out of the spaceships. However, the tossing of these clocks from the spaceships does not suddenly cause an increase in the length of spaceship X and the elimination of its .866 contraction with respect to observers in spaceship Y. In like manner, the tossing out of these clocks does not cause the rate of time in spaceship X to suddenly be the same as the rate of time in spaceship Y relative to the spaceship Y observers. In other words, the tossing of these clocks from the spaceships causes no change concerning the time dilation and length contraction of spaceship X with respect to observers in spaceship Y. The reason that the tossing of the clocks causes no changes from the way things were before their tossing is that the clocks play no role in causing the time dilation and length contraction that were occurring when they were in the spaceships. The differing rates of time and the differing lengths of spaceship X that took place with respect to the personnel in spaceship X and spaceship Y are results caused by the two principles of special relativity functioning through the .5c uniform motion between spaceship X and spaceship Y. Clocks can reveal various details that take place when time dilation and length contraction are created. However, the clocks play no role in the creation of time dilation and length contraction and these phenomena will occur with or without the clocks.

Concerning the experiment in this chapter, it will take place in accord with the claims and principles of special relativity. However, what happens in the experiment will require no revelations from clocks, and no clocks will be used in the experiment.

The Experiment

Section A

The Firing of Pulse One over an A to AA to A Vertical Length

The experiment begins like the one at the beginning of this chapter. Hence, spaceship X has a length of 100 feet as measured by its occupants. Also as measured by the occupants, point A is on the floor and point AA is 16 feet vertically above point A on the ceiling. There is also motion between spaceship X and spaceship Y at a speed of .5c. When a pulse of light is fired in spaceship X from a projector at point A on the floor, it travels from A up to AA and from AA down to A for a total distance of 32 feet with respect to the spaceship X occupants. And on the basis that the constant speed of light is one-foot-per-nanosecond, the pulse has an elapsed time of 32 nanoseconds.

With respect to the observers in spaceship Y, the .5c motion of spaceship X and the principles of special relativity cause the journey of the pulse from A to AA to A to increase over the 32 feet that it would have without this motion by a factor of .866. Thus the pulse has an elapse time of 36.95 nanoseconds (32 ÷ .866 = 36.95) in its A to AA to A journey with respect to the observers in spaceship Y. Furthermore, the pulse forms the sides of an isosceles triangle.

Hence, by the pulse in spaceship X having an elapsed time of 32 nanoseconds with respect to the occupants of spaceship X and by this same pulse having an elapsed time of 36.95 nanoseconds with respect to the observers in spaceship

Y, the pulse is revealing that the rate of time in spaceship X is slower than the rate of time in spaceship Y by a factor of .866 (32 ÷ 36.95 = .866) relative to the spaceship Y observers.

Concerning the length of spaceship X with respect to the spaceship Y observers, the dilation of time in spaceship X by a factor of .866 causes the length of spaceship X to contract by that same .866 factor. Hence, the 100 foot length of spaceship X as measured by its occupants is contracted to a length of 86.6 feet with respect to the spaceship Y observers (100 x .866 = 86.6).

Section B

The Firing of Pulse Two over an A to B to C Isosceles Length

The projector at point A in spaceship X is designed to split the ray from the projector into two separate pulses of light, pulse one and pulse two. With respect to the occupants of spaceship X, pulse one is the returned pulse that travels 16 feet from point A on the floor to point AA that is vertically above point A and then returns to point A for an elapsed time of 32 nanoseconds.

Concerning the journey of pulse two in spaceship X with respect to the occupants of spaceship X, the pulse travels diagonally from the projector at point A on the floor to point B on the ceiling that is behind point AA on the ceiling. The length from point AA to point B is measured at 7.28 feet by the spaceship X occupants. The pulse then travels diagonally from point B to point C on the floor that is 7.28 feet behind point B on the ceiling and 14.56 feet behind point A on the floor. Hence, in emerging from point A, pulse two forms the two sides of an isosceles triangle as it travels toward the back of spaceship X and against any motion of the spaceship. As to the length of these two diagonal sides, each of them is the hypotenuse of a right triangle with a height of 16 feet and a base of 7.28 feet. Hence, in accord with the Pythagorean Theorem, each of these hypotenuses has a length of 17.57835 feet. Thus, with respect to the occupants of spaceship X, pulse two travels over a total length of 35.1567 feet (17.57835 x 2 = 35.1567) in 35.1567 nanoseconds.

As to what takes place concerning pulse two with respect to the observers in spaceship Y and in accord with the claims of special relativity, the .5c motion of spaceship X causes point B on the ceiling to move forward as pulse two heads toward point B from point A on the floor. This forward motion of point B causes a decrease in the distance and elapsed of the pulse from what they would be without this .5c motion with respect to the spaceship Y observers. In like manner, the .5c motion of spaceship X causes point C on the floor to move forward as pulse two heads toward it from point B on the ceiling. This also causes a decrease in the distance and elapsed of the pulse from what they would be without this motion. This decreased elapsed time of pulse two in its A to B to C journey means that, with respect to the spaceship Y observers, the motion of spaceship X is causing the rate of time in spaceship X to be faster than in spaceship Y even as the increased elapsed time of pulse one meant that the motion of spaceship X was causing the rate of time in spaceship X to be slower than in spaceship Y. And by this .5c motion of spaceship X causing an increase in the rate of time in spaceship X with respect to the spaceship Y observers, this motion of spaceship X is also causing an increase in the length of spaceship X and every parallel length in the spaceship. Furthermore, the factor by which the rate of time is increased must also be the factor by which the length of spaceship X and every parallel length in spaceship X are increased. This common factor is .91, an amount determined by the trial-and-error method.

Hence, on the basis of a .91 factor, the 100 foot length of spaceship X as measured by its occupants is increased to a length of 109.89 feet (100 ÷ .91 = 109.89) with respect to the spaceship Y observers.

We then consider the distance between point AA and point B on the ceiling of spaceship X that has a length of 7.28 feet as measured by the spaceship X occupants. By being increased by a factor of .91 with respect to the observers in spaceship Y, the point AA to point B length is now 8 feet (7.28 ÷ .91 = 8). Hence, when pulse two is fired from point A on the floor of spaceship X to point B on the ceiling, the .5c speed of spaceship X with respect to the spaceship Y observers causes points AA and B on the ceiling to move 8 feet forward in 16 nanoseconds and causes point B to be directly over where point A had been when pulse two was fired from point A. During these same 16 nanoseconds, pulse two from point A travels up to point B on a vertical pathway of 16 feet and reaches point B. Pulse two then returns to point C with an elapsed time of 32 nanoseconds with respect to the spaceship Y observers.

Thus, by pulse two having an elapsed time of 35.1567 nanoseconds with respect to the spaceship X occupants and an elapsed time of 32 nanoseconds with respect to the space-ship Y observers, the pulse is revealing that the rate time in spaceship X is faster than the rate of time in spaceship Y by a factor of .91 (32 ÷ 35.1567 = .91) with respect to the spaceship Y observers. For every nanosecond that passes in spaceship X, .91 nanosecond passes in the spaceship Y. And as shown previously, the 7.28 foot length between point AA and B on the ceiling as measured by spaceship X occupants is increased by a factor of .91 to 8 feet with respect to the spaceship Y observers, and the 100 foot length of spaceship X as measured by its occupants is increased by a factor of .91 to a length of 109.89 feet (100 ÷ .91 = 109.89) with respect to the spaceship Y observers.

Section C

The Claims of Special Relativity and Their Impossible Results

Impossible result one. Through the use of pulse one in section A, it was shown in the experiment how the claims of special relativity and the uniform motion between spaceship X and spaceship Y cause the rate of time in spaceship X to be slower than the rate of time in spaceship Y by a factor of .866 with respect to the spaceship Y observers. Then through the use of pulse two in section B, it was shown in the previous experiment how the claims of special relativity and the uniform motion between the two systems cause the rate of time in spaceship X to be faster than the rate of time in spaceship Y by a factor of .91 with respect to the spaceship Y observers. It is, of course, totally impossible that, with respect to the observers in spaceship Y, time in spaceship X is passing slower than in spaceship Y while time in spaceship X is passing faster than in spaceship Y. And by the claims of special relativity bringing about such impossible results, the claims of special relativity are negated.

Impossible result two. Through the use of pulse one in section A, the experiment shows how the claims of special relativity and the uniform between spaceship X and spaceship Y cause the length of spaceship X to be contracted by a factor of .866 with respect to the observers in spaceship Y. Hence, the 100 foot length of spaceship X as measured by its occupants was contracted to a length of 86.6 feet with respect to the spaceship Y observers. Then through the use of pulse two in section B, the experiment shows how the claims of special relativity and the uniform motion between the two systems cause the length of spaceship X to be increased by a factor of .91 with respect to the observers in spaceship Y. Hence, the 100 foot length of spaceship X as measured by its occupants was increased to a length of 109.89 feet with respect to the spaceship Y observers. Thus, with respect to the spaceship Y observers, we have the impossible result of spaceship X having a length of 86.6 feet while having a length of 109.89 feet. And by the claims of special relativity bringing about these impossible results, the claims of special relativity are again negated.

Added Note

In the experiment of this chapter, there was uniform motion between spaceship X and spaceship Y at a speed of .5*c*. In spaceship X, point A was on the floor of the spaceship, point B was behind point A on the ceiling, and point C was behind point B on the floor. Hence, a line from A to B to C forms the sides of an isosceles triangle. It was then shown that, with respect to the spaceship X occupants, the elapsed time of pulse two from A to B to C was 35.1567 nanoseconds. It was also revealed that, with respect to the observers in spaceship Y, the .5c motion between the spaceships and the principles of special relativity caused pulse two to have an elapsed time of 32 nanoseconds. However, there were no clocks in spaceship X or Y to reveal what was caused by the .5c motion and the principles because they were not needed. The way in which the experiment was structured and the all-important determination of the .91 factor concerning length contraction and time dilation enabled the experiment to reveal its results without the help of clocks. However, if the claim is made that the elapsed time of pulse two over the A to B to C isosceles pathway must be revealed by synchronized clocks at points A and C in order for this experiment to be valid, then this claim can readily be accommodated. It was shown in the last chapter by the Revised Synchronization Approach that during the motion between spaceship X and Y and in accord with the claims of special relativity, clocks in spaceship X can be fully synchronized with respect to the occupants of spaceship X as well as with respect to the observers in spaceship Y.

CHAPTER 6

Evidence Refuting the Claims of Special Relativity From the Darkness and Light in a Spaceship

The experiment in this chapter will deal with a windowless spaceship, spaceship X, in which there is total darkness unless a beam of light is fired from a source at the back of the spaceship that can bring light over the entire length of the spaceship. The experiment will then show how the claims of special relativity can invalidate themselves by causing darkness to fill a large part of the spaceship with respect to the observers in spaceship Y while light fills the entire spaceship relative to the occupants of spaceship X.

Overview of the Experiment and a Question

The experiment begins with uniform motion between spaceship X and spaceship Y at a speed of .9999c. In accord with the claims of special relativity, none of this motion between the two spaceships is the motion of spaceship X with respect to the occupants of spaceship X. Also with respect to these occupants, spaceship X has a length of a hundred feet from point A at the back of the spaceship to point B at the front of the ship as measured by the occupants. Furthermore, unless a continuous beam of light is fired from A to B, the windowless spaceship is in total darkness. However, when a beam is sent from point A, it brings light throughout spaceship X in 100 nanoseconds with respect to its occupants on the basis that the speed of light is one-foot-per-nanosecond. The light continues to fill the spaceship until the beam is turned off.

Relative, however, to the spaceship Y observers, the .9999c motion between spaceship X and spaceship Y is entirely the motion of spaceship X. Hence, when the beam is fired from point A to B in spaceship X, point B moves away from the beam at a speed of .9999c, almost the speed of light, because of the speed of spaceship X and the source of the beam in spaceship X not being added to the speed of the beam in accord with the constant speed of light principle. That then brings us to this question: With respect to the spaceship Y observers, how much time must pass in order for a beam of light to travel from point A to point B in spaceship X and bring light throughout the spaceship while point B is moving away from the beam almost at the velocity of the beam?

Answering the Question

In order to answer this question, we will have a beam not just traveling from A to B in spaceship X, but a beam that travels from A to B and is reflected from B back to A. However, the beam will still be traveling from A to B, and it can still be used to reveal how much time must pass in order for the beam to travel from A to B. Hence, with respect to the occupants of spaceship X, the returned beam travels 100 feet from A to B and 100 feet from B to A for a total elapsed time of 200 nanoseconds.

Concerning the elapsed time of this returned beam with respect to the observers in spaceship Y, the claims and principles of special relativity and the .9999c motion of spaceship X cause time in spaceship X to pass slower by a certain factor than what it would be without this motion. The claims of special relativity and the .9999c motion of spaceship X also cause a contraction in the A to B length of spaceship X from the 100 foot length that it would have without this motion. Furthermore, the factor by which time in spaceship X is passing slower is the same factor by which the length of the spaceship is contracted. As will be shown in what follows, this common factor is .014, a factor determined by the use of the Lorentz-Fitzgerald contraction, $\sqrt{1-v^2/c^2}$. Hence, by v being 0.9999c and c being 1.00c, the contraction reads $\sqrt{1-.9999^2 \div 1.00^2}$ and gives us a factor of .014.

Thus, by the claims of special relativity and the .9999c speed of spaceship X relative to the spaceship Y observers causing the 100 foot A to B length of spaceship X to be contracted by a factor of .014, spaceship X has a shortened length of 1.4 feet (100 x .014 = 1.4). The returned beam is then fired over the 1.4 foot length from A to B and the 1.4 foot length from B to A. Concerning the elapsed time of the beam in its journey from A to B, the .9999c speed of spaceship X causes point B to move away from the beam at a speed of .9999c in accord with the constant speed of light principle. This motion of B away from the beam causes an increase in the elapsed time of the beam over what it would with without this motion. The amount of this increased elapsed time is the

1.4 foot contracted length of the spaceship being divided by c - v (the one-foot-per-nanosecond speed of the beam minus the .9999c speed of the spaceship). This gives us an elapsed time of 14,000 nanoseconds for the A to B segment of the beam (1.0000c - .9999c = .0001 and 1.4 ÷ .0001 = 14,000 nanoseconds). Concerning the elapsed time of the beam in its journey from B to A, the .9999c speed of the spaceship causes point A to move toward the beam at a speed of .9999c in accord with the constant speed of light principle. This motion of point A toward the beam causes a decrease in the elapsed time of the beam from what it would with without this motion. The amount of this decreased time is the 1.4 foot contracted length of the spaceship being divided by c + v (the one-foot-per-nanosecond speed of the beam plus the .9999c speed of the spaceship). This gives us an elapsed time of .7 nanosecond for the B to A segment of the beam (1.00c + .9999c = 1.9999 and 1.4 ÷ 1.9999 = .7). Thus, when these two elapsed times are added together, the entire beam has an elapsed time of 14,000.7 nanoseconds with respect to the spaceship Y observers.

Thus, as shown, the beam in spaceship X has an elapsed time of 200 nanoseconds in it's A to B to A journey with respect to the spaceship occupants and an elapsed time of 14,000.7 nanoseconds with respect to the observers in spaceship Y. Thus, the beam is revealing that, with respect to the spaceship Y observers, the .9999c motion of spaceship X is causing time in the spaceship to pass slower than without this motion by a factor of .041 (200 ÷ 14,000.7 = .041). It was also shown that, relative to the spaceship Y observers, the .9999c motion of spaceship X caused the length of spaceship X to be contracted by a factor of .041. Hence, with respect to the spaceship Y observers, even as the length of spaceship X is contracted by a factor of .041, so also is the speed of time in spaceship X dilated by a factor of .041.

At this point we are ready to return to the question we previously asked: "With respect to the spaceship Y observers, how much time must pass in order for a beam of light to travel from point A to point B in spaceship X and bring light throughout the spaceship while point B is moving away from the beam almost at the velocity of the beam?" As shown, when the returned beam in spaceship X is fired from A to B to A during the .9999c motion between spaceship X and spaceship Y, it has an total elapsed time of 14,000.7 nanoseconds with respect to the spaceship Y observers. The journey of the beam from A to B has an elapsed time of 14,000 nanoseconds, and its journey from B to A has an elapsed time of .7 nanosecond. Hence, the answer to the question is that 14,000 nanoseconds must pass in order for the beam in spaceship X to travel from A to B and bring light throughout the spaceship with respect to the observers in spaceship Y.

It was also shown that, with respect to the spaceship X occupants, it takes 200 nanoseconds for the beam to travel from A to B to A, and it takes 100 nanoseconds for the beam to travel from A to B.

An Impossible Result Created by the Claims of Special Relativity

In the experiment of this chapter, there is uniform motion between spaceship X and spaceship Y at a speed of .9999c. Also in the experiment, spaceship X is without windows, and this causes the spaceship to be in total darkness. A light beam is then fired from point A at the back of spaceship X that travels to point B at the front of the spaceship in order to bring light throughout this A to B length. The experiment then considered how much time it will take the beam in spaceship X to travel from A to B with respect to the occupants of spaceship X and the observers in spaceship Y in accord with the claims of special relativity.

With respect to the occupants of spaceship X, the experiment shows that it takes 100 nanoseconds for the beam to travel from A to B and bring light throughout the spaceship X. However, with respect to observers in the spaceship Y, it takes 14,000 nanoseconds for the beam to bring light over the entire A to B length of spaceship X. Hence, with respect to the spaceship Y observers, in 100 nanoseconds, light has come to only .7 percent of the length of spaceship X (100 ÷ 14,000 = .007). The other 99.3 percent of spaceship X is in darkness relative to the spaceship Y observers. Furthermore, this darkness is complete darkness since not a single photon can move faster than c. However, it is impossible for light to be filling spaceship X with respect to the occupants of spaceship X while 99.3 percent of spaceship is in complete darkness relative to the spaceship Y observers. Thus, by the claims of special relativity bringing about this impossible result, these claims are shown to be erroneous and invalid.

Chapter 7

Evidence Refuting the Claims of Special Relativity From Special Relativity's Length Contraction

According to the claims of special relativity, when spaceship X is in uniform motion relative to observers in spaceship Y, this motion of spaceship X causes time in spaceship X to pass slower than before this motion by a certain factor. Also with respect to the spaceship Y observers, the motion of spaceship X causes the length of spaceship X and the length of everything in spaceship X that is parallel to its motion to be contracted from what it was before this motion by a factor that is the same as the slowing of time factor. Two experiments in this chapter will demonstrate how special relativity's length contraction can invalidate the claims of special relativity. And as in all the chapters of this book, one-foot-per-nanosecond will be used as the constant speed of light.

Experiment One

Experiment one will be conducted in accord with the claims of special relativity and will test these claims in a very dramatic way. The experiment begins with uniform motion between spaceship X and spaceship Y at a speed of .968c.

Spaceship X with Respect to Its Occupants

With respect to the occupants of spaceship X, the spaceship has a length of 60 feet as measured by the occupants, and within the spaceship is a room that has a width and length of ten feet as also measured by the occupants. Wall F of the room is the wall closest to the front end the spaceship, and wall B is the wall that is closest to the back end of the spaceship. Both walls have a height of 8 feet as measured by the occupants and extend from one side of the spaceship to the other side. As indicated, wall B is ten feet behind wall F.

Within the room are two ropes that can travel through the room on different pathways. One of the ropes, rope c (the ceiling rope), enters the room from an opening at the top of wall B, the back wall, by passing over a pulley at this opening. From wall B, rope c extends for a length of 10 feet over the ceiling of the room to a pulley on wall F, the front wall. From the pulley at the top of wall F, rope c again extends over the ceiling for a length of ten feet back to wall B. Rope c then passes out of the room though the opening at the top of wall B over a second pulley at this opening. The pulleys at the opening in the top of wall B and the pulley at the top of wall F are imbedded in the walls in such a way so that the journey of rope c over the ceiling has a total distance of twenty feet. Another feature concerning rope c is that a clip is attached to a point on the rope that is at the opening on the top of wall B. Hence, when rope c travels twenty feet over the ceiling from wall B to wall F to wall B, the clip will be traveling twenty feet with rope c. Then when the clip has returned to wall B and passes through the opening at the top of wall B, the clip will hit switch S. Furthermore, if the switch is not disabled, the switch will cause explosives to ignite and cause the spaceship to be destroyed. Thus, the clip on rope c is similar to the clip on the chain of a garage door opener that can throw a switch and cause the garage door to open or close.

The second rope, rope w (the wall rope), also enters the room from the opening at the top of wall B by passing over a third pulley at this opening. From the opening, rope w extends over the 8 foot height of wall B to a pulley at the bottom of the wall where the wall and the floor meet. From the pulley at the bottom of the wall, rope w again extends over the 8 foot height of wall B to the opening at the top of wall B, where it passes out of the room on a fourth pulley at this opening. The pulleys at the opening in the top of wall B and the pulley at the bottom of the wall are imbedded in wall B in such way so that within the room, rope w travels a total distance of 16 feet. As was the case with rope c, a clip is attached to a point on rope w that is at the opening on the top of wall B. Hence, when rope w travels 16 feet from the top to the bottom to the top of wall B, the clip will be traveling these 16 feet with rope w. Then when the clip travels through the opening at the top of wall F, it will hit switch S. And if the switch is still active, the clip will cause it to be disabled. Switch S at this opening is located between ropes w and c in such a way that it can be touched by the clip on each of these ropes.

As to how ropes c and w are able to move through the room from the opening at the top of wall B and back to this opening, the two ropes are wrapped around a non-motorized spool in an area behind wall B, the back wall, and from this spool they enter the opening at the top of wall B. The two ropes can then travel through the room by both ropes being attached to a motorized spool that is also in the area behind wall B. When turned on, the motorized spool can pull ropes c and w through the room. Hence, by the two ropes being attached to the same motorized spool, they travel through the room at the same speed. This common speed is one-foot-per-second as determined by the spaceship X occupants.

The Motion of Ropes w and c
With Respect to the Occupants of Spaceship X

Having presented the rope setup in the spaceship X room, we continue with the motorized spool in spaceship X pulling ropes w and c through the room with respect to the occupants of spaceship X. The clip on rope w travels from the opening at the top of wall B over a 16 foot pathway from the top to the bottom to the top of wall B. By traveling at a speed of one-foot-per-second, it takes 16 seconds for the clip on rope w to hit switch S at the top of wall B and cause it to be disabled. Hence, we have this equation concerning the clip on rope w: *16 foot length ÷ one-foot-per-second speed = 16 second elapsed time.*

As to the clip on rope c, the clip that can cause spaceship X to be destroyed, it travels over the ceiling from the opening at the top of wall B over a 20 foot pathway from wall B to wall F to wall B. By traveling at a speed of one-foot-per-second, it takes 20 seconds for the clip on rope c to hit switch S. Thus, the equation concerning the clip on rope c has this reading: *20 foot length ÷ one-foot-per-second speed = 20 second elapsed time.* And by the clip on rope w reaching switch S 4 seconds before the clip on rope c can reach the switch, switch S cannot cause the explosives to ignite because the switch had been disabled by the clip on rope w. Thus, no harm comes to spaceship X with respect to the occupants of spaceship X.

Another event that takes place somewhere in spaceship X is the firing of a pulse from point one on the spaceship floor to point two on the ceiling, which is 8 feet vertically above point one. The pulse is then reflected back to point one. With respect to the occupants of spaceship X, the pulse travels a total distance of 16 feet and has an elapsed time of 16 nanoseconds.

The Motion of Ropes w and c With Respect to the Observers in Spaceship Y

Having considered what takes place concerning spaceship X with respect to the occupants of spaceship X during the .968*c* motion between the two systems, we now consider what takes place concerning spaceship X relative to the observers in spaceship Y.

We begin by pointing out that, in accord with the principles of special relativity, all of the motion between spaceship X and spaceship Y is the motion of spaceship X with respect to the spaceship Y observers. Hence, when the pulse is fired over the 16 foot vertical length from point one to two to one, the .968*c* motion of spaceship X causes an increase in the elapsed time of this pulse. The amount of this increase can be determined by dividing the 16 foot vertical length over which the pulse moves by the Lorentz-Fitzgerald contraction, $\sqrt{1-v^2/c^2}$. Hence, by the velocity of the spaceship being .968c and by the constant speed of light being 1.0c, the Lorentz-Fitzgerald contraction reads .25. This means that, relative to the spaceship Y observers, the pulse in spaceship X has an elapsed time of 64 nanoseconds (16 ÷ .25 = 64). Hence, the spaceship X pulse has an elapsed time of 16 nanoseconds relative to the occupants of spaceship X and an elapsed time of 64 nanoseconds relative to the observers in spaceship Y. This means that, with respect to the spaceship Y observers, the .968c speed of spaceship X is causing the rate of time in the spaceship to be slower than what it would be without this motion by a factor of .25 (16 ÷ 64 = .25). For every 4 nanoseconds that would have passed in spaceship X before this motion, only one nanosecond passes during the motion.

Furthermore, by the .968c speed of spaceship X causing a slowing of time in spaceship X by a factor of .25 relative to the spaceship Y observers, this slowing of time in spaceship X by a factor of .25 also means that the length of spaceship X and any length inside the spaceship that is parallel to the motion of the spaceship is contracted by this same .25 factor. Hence, with respect to the spaceship Y observers, the 60 foot length of spaceship X as measured by its occupants is contracted to fifteen feet (60 x .25 = 15), and the ten foot length of the spaceship room as measured by the occupants is contracted to 2.5 feet (10 x .25 = 2.5).

The experiment then continues with the journeys of ropes c and w through the spaceship X room during the .968c speed of spaceship X with respect to the spaceship Y observers.

Journey of Rope c
With Respect to the Spaceship Y Observers

We first consider the journey of rope c and its clip over the spaceship ceiling from wall B (back wall) to wall F (front wall) and back to wall B. We begin by pointing out that the .968c speed of spaceship X with respect to the spaceship Y observers causes the ten foot length of the ceiling to contract to a length of 2.5 feet (10 x .25 = 2.5 feet) by this length being parallel to the motion of spaceship X. Hence, the length over which the clip on rope c travels in its journey over the ceiling from wall B to wall F to wall B is 5 feet (2.5 + 2.5 = 5).

Thus, if there were only a length contraction of spaceship X and no time dilation in the spaceship, the clip on rope c would travel these 5 feet at a speed of one-foot-per-second and have an elapsed time of 5 seconds for its two-way journey over the ceiling as shown by this equation: *5 foot length ÷ one-foot-per-second speed = 5 second elapsed time*. However, by time in spaceship X dilating by a factor of .25, the elapsed time of the clip on rope c is not 5 seconds but 1.25 seconds (5 x .25 = 1.25). And by the clip on rope c traveling over a length of 5 feet in 1.25 seconds, this means that the motion of spaceship X with respect to the observers in spaceship Y is causing the speed of the clip to increase from one-foot-per-second to 4-feet-per-second (*5 foot length ÷ 1.25 second elapsed time = 4-foot-per-second speed*).

Journey of Rope w
With Respect to the Spaceship Y Observers

We then continue with what takes place concerning the journey of the clip on rope w over wall B. The journey begins at the top of wall B, the back wall, and continues with the clip on rope w moving down to the bottom of the wall and then moving up to the top of the wall. Since wall B is at a right angle relative to the motion of the spaceship X, there is no contraction of the wall's 8 foot height with respect to the spaceship Y observers. Hence, the clip on rope w travels over a 16 foot length by moving from the top to the bottom to the top of wall B. Thus, if there were no time dilation in the spaceship, the clip on rope w would travel over this 16 foot length at a speed of one-foot-per-second and have an elapsed time of 16 seconds for its journey over wall B in accord with this equation: *16 foot length ÷ one-foot-per-second speed = 16 second elapsed time.* However, by time in the spaceship dilating by a factor of .25, the elapsed time of the clip on rope c is not 16 seconds but 4 seconds (16 seconds x .25 = 4 seconds). And by the clip on rope c traveling over a length of 16 feet in 4 seconds, this means that the motion of the spaceship with respect to the spaceship Y observers has caused the speed of the clip to increase from one-foot-per-second to 4-feet-per-second (*16 foot length ÷ 4 second elapsed time = 4-foot-per-second speed*).

Impossible Results by the Claims of Special Relativity

Thus, with respect to the observers in spaceship Y and as shown in the previous two paragraphs, the .968c speed of spaceship X and the claims of special relativity cause the clip on rope c to complete its two-way journey over the spaceship ceiling in 1.25 seconds and cause the clip on rope w to complete its two-way journey over the back wall of the spaceship in 4 seconds. Hence, 2.75 seconds before the clip on rope w can disable the explosive switch, the clip on rope c causes the switch to be thrown, the explosives to be ignited, and the spaceship to be destroyed with respect to the spaceship Y observers.

However, it was also shown that, with respect to the spaceship X occupants and in accord with the claims of special relativity, the clip on rope c completes its two-way journey over the spaceship ceiling in 20 seconds, and the clip on rope w completes its two-way journey over wall B in 16 seconds. Hence, by the clip on rope w reaching the switch and causing it to be disabled 4 second before the clip on rope c can cause the switch to be thrown, no harm comes to the spaceship relative to the spaceship occupants.

Hence, the experiment is revealing how the claims of special relativity create the impossible result of a spaceship that is destroyed with respect to observers in spaceship Y, while remaining whole with respect to occupants in spaceship X. And by the claims of special relativity creating such an impossible result, the claims of special relativity are nullified.

Experiment Two

Experiment two will be conducted in accord with the claims of special relativity and will continue with uniform motion between spaceship X and spaceship Y at a speed of .968c as in experiment one.

However, unlike experiment one, there is no special room in spaceship X through which ropes are pulled in experiment two. But there is a square box in spaceship X with a height, width, and length of 4 feet with respect the occupants of spaceship X. Concerning the 4 sides of the box, the front side and the back side are vertical to the length of spaceship X and any motion of spaceship X. The other two sides are parallel to the length of the spaceship and to any motion of the spaceship.

As shown in experiment one, the .968c speed of spaceship X with respect to the spaceship Y observers and the claims of special relativity cause the length of spaceship X and every length in spaceship X that is parallel to the spaceship's motion to contract by a factor of .25 from what that length would be without this motion. Thus, we return to the box in spaceship X that has a height, width, and length of 4 feet with respect the occupants of spaceship X.

With respect to the observers in spaceship Y, the box continues to have a width of 4 feet and height of 4 feet because the width and height of the box are not parallel to the motion spaceship X. However, the 4 foot length of the box is parallel to the motion of spaceship X. Hence, this length is contracted by a factor of .25 to a length of one foot. Thus, with respect to the spaceship Y observers, the box has a height that is 4 times greater than its length.

Thus, during the .968c motion between spaceship X and spaceship Y, there is a single box in spaceship X. The claim then comes from special relativity through the functioning of its principles that, with respect to the spaceship X occupants, this single box has a height that is the same as its length and that there is no box in the spaceship with a height that is 4 times greater than its length. It is also claimed by special relativity that, with respect to the spaceship Y observers, this single box in the spaceship has a height that is 4 time greater than its length and that there is no box in the spaceship with a height that is the same as its length. But can this claim of special relativity be factual? In what follows, this claim will be tested.

While there is motion between the spaceship X and spaceship Y at a speed of .968c, an occupant of spaceship X stands midway between the front and back of the box and a few feet away from the box. She proceeds to take a picture of the box with her camera. Then when she and the other occupants of spaceship X physically have the picture in their hands, they see on this picture the image of a box with a height that is the same as its length. And what the spaceship occupants see on the picture is in complete accord with what is claimed by special relativity.

After the occupants of spaceship X physically held and looked at the picture, they send the picture to spaceship Y by email. It is then the observers in spaceship Y that physically have the picture in their hands and, in viewing it, they see on the picture the image of a box with a height and length that are the same. The image of the box that the occupants of spaceship X saw on the picture when they emailed the picture to the spaceship Y observers must be the image of the box that the spaceship Y observers see on the picture when the emailed picture arrives at the spaceship Y. Hence, even as the occupants of spaceship X saw on this picture a box with a height and length that are the same, the occupants of spaceship Y must also see on this picture a box with a height and length that are the same.

Furthermore, the picture of the box that the spaceship Y observers are physically seeing was taken by a camera in spaceship X. And since a camera can only take a picture of what is in front of it, it must be that in front of this camera in spaceship X is a box with a height that is the same as its length. Hence, by the spaceship Y observers physically seeing a picture taken by this camera, the picture is showing them that the box in front of the camera has a height that is the same as its length.

Thus, the previous claim by special relativity that there is no box in spaceship X with a height that is the same as its length with respect to the spaceship Y observers is shown to be erroneous. And by making this erroneous assertion, the claims of special relativity are again revealing their lack of validity.

Furthermore, by the box in spaceship X having a height that is the same as its length with respect to the spaceship X occupants and the spaceship Y observers during the .968c motion between the two spaceships, this means that this motion and the claims of special relativity cause no length contraction of the box with respect to the personnel in these two systems. And by the principles causing no length contraction of the spaceship box, they also are causing no contraction of the spaceship. In like manner, by there being no length contraction of the spaceship, there also is no time dilation in the spaceship because of the way these two results are linked together. Hence, the picture of the spaceship box is revealing that the time dilation and length contraction claims of special relativity do not occur. Thus, the claim of special relativity that the functioning of its principles can bring about time dilation and length is shown to be erroneous and invalid.

CHAPTER 8

Evidence Refuting the Claims of Special Relativity from a Pulse Fired on a Diagonal Pathway

Introduction to the Experiment

The experiment in this chapter will be conducted in accord with the claims and principles of special relativity and will be using one-foot-per-nanosecond as the constant speed of light. In the experiment, point A is on the floor of a spaceship and point B is on the spaceship's ceiling. However, point B is behind point A. When a pulse of light is fired from point A, it travels diagonally back to point B. At point B, the pulse hits a mirror that is at a right angle relative to the pulse, and this causes the pulse to return to point A. Hence, with respect to the occupants of the spaceship, the pathway of the pulse from A to B is also the pathway of the pulse from point B to A. There is no angle between these two pathways. With respect, however, to observers before whom the spaceship is in motion and in accord with the claims of special relativity, point B moves forward as the beam travels from A to B. This means that when the pulse from A hits the mirror at point B, the mirror is not at a right angle to the pulse. Hence, the pathway of the pulse from the mirror at point B is at an angle with respect to the pathway of the pulse to the mirror at point B. Thus, there is no angle between the pathway of the pulse to B and the pathway from B with respect to occupants of the spaceship, but there is an angle between these two pathways with respect to observers before whom the spaceship is in motion. The experiment in this chapter will then show how these different pathways with respect to the spaceship's occupants and observers can create impossible results.

The Experiment

The experiment begins with uniform motion between spaceship X and spaceship Y at a speed of $.593c$. Within spaceship X, point A is on the floor of the spaceship, and point B is on the spaceship's ceiling behind point A. The vertical length between the floor and the ceiling as well as between points A and B is 13.5 feet. With respect to the occupants of spaceship X, when a pulse of light is fired from a source at point A, the pulse travels from point A diagonally back to point B, where it hits a mirror. The mirror is at a right angle relative to the pulse and sends the pulse from point B forward to point A on the floor. Hence, the pathway of the pulse from the mirror at point B to point A is also the pathway of the pulse from point A to B.

Between the ceiling and the floor of spaceship X is a wall to wall platform that covers the entire front end of the spaceship. Thus, where the platform is located, it is as though spaceship X is a structure with two stories. Square photoelectric cells cover the entire surface of the platform, and the distance between the platform cells and the ceiling is 9 feet. Explosives are attached to the entire underside of the platform, and if any one of the photoelectric cells is hit by a pulse of light, then the cell will cause the explosives to ignite and the spaceship to be destroyed. However, with respect to the occupants of spaceship X, when the pulse travels diagonally from A to B and diagonally from B to A, it passes next to the end of the platform but does not touch it. Thus, no harm comes to spaceship X with respect to its occupants.

We then consider what happens in spaceship X with respect to the observers in spaceship Y during .593c speed between the two systems. Relative to the spaceship Y observers and in accord with the relativity principle of special relativity, all of this motion between the spaceships is the motion of spaceship X. Furthermore, this motion of spaceship X and the claims of special relativity cause time in the spaceship to pass slower than what it would be without this motion. They also cause a contraction in the length of spaceship X and every length in the spaceship that is parallel to this motion. Hence, we will say that the length of the platform in spaceship X has been contracted to 100 feet. We will also say that the length between A and B (the distance that B on the ceiling is behind A on the floor) has been contracted to a length of 8 feet.

We continue with the impact on the pulse fired from point A in spaceship X with respect to the spaceship Y observers only on the basis that the length between A and B has been contracted to a length of 8 feet without considering any additional impact on the pulse by the spaceship's motion. We point out that this contraction changes the angle of the pulse relative to the ceiling and the floor and changes the two-way distance of the pulse from what it would be without this contraction. In a contracted length, things are squeezed together, and this can affect their lengths and angles. However, if a pulse travels from A to B to A before length contraction, it will travel from A to B to A during length contraction but with changed angles and distances. By way of comparison, if a rod extended diagonally from A to B in spaceship X, the contraction of the rod would change the angle and length of the rod, but the rod would continue to extend from A to B.

Hence, by the vertical length between point A on the floor and point B on the ceiling being 13.5 feet and by the parallel length between points A on the floor and point B on the ceiling being contracted to 8 feet, the diagonal length from point A to point B is the hypotenuse of a right triangle with an 8 foot base and a 13.5 foot height. Thus, in accord with the Pythagorean Theorem, the A to B length is 15.7 feet. This means that, if the principles of special relativity caused only length contraction, the pulse would travel 15.7 feet from A to B and 15.7 feet from B to A for a total distance of 31.4 feet. And, in order for the pulse to return from B to A, the mirror on the ceiling must continue to be at a right angle relative to the pulse. Furthermore, by the pulse having a length of 15.7 feet in traveling from A on the floor to B on the ceiling, and by the distance between the floor and the ceiling being 13.5 feet, it can be determined by this information and by the use of a protractor that, when the pulse arrives at point B on the ceiling, it is at an angle of 120° relative to the ceiling that is between point B and the back of the spaceship. And by the mirror at point B being at a 90° angle relative to the pulse, the mirror is at an angle of 30° with respect to the ceiling that is between point B and the back of the spaceship (120° - 90°

= 30°). This 30° angle of the mirror will remain throughout the .593c motion between the two spaceships.

Thus, having considered the impact of length contraction on the pulse, we continue with the full impact of motion and the principles of special relativity on the pulse in spaceship X relative to the observers in spaceship Y. Hence, when the pulse is fired from point A on the floor, it begins its journey to point B on the ceiling that is 8 feet behind point A. While the pulse is traveling to point B, the motion of spaceship X relative to spaceship Y at a speed of .593 feet per nanosecond causes point B to move 8 feet forward in 13.5 nanoseconds (13.5 x .593 = 8), and hence, point B is vertically over where the pulse began its journey from point A. During these same 13.5 nanoseconds, the pulse from point A travels vertically to where point B is now located, a distance of 13.5 feet, and hits the mirror. And since the pulse is traveling vertically to point B on the ceiling, the angle between the ceiling and the pulse is now 90°.

Furthermore, as previously pointed out, the mirror at point B has an angle of 30° relative to the ceiling that is between point B and the back of the spaceship. Hence, by the mirror having an angle of 30° relative to the ceiling and by the pulse having an angle of 90° relative to the ceiling, there must be an angle of 60° between the mirror and the incoming pulse (90° - 30° = 60°).

Concerning the outgoing pulse, it must leave the mirror at an angle of 60° even as there is an angle of 60° between the mirror and the incoming pulse. Since the incoming pulse has an angle of 90° relative to the ceiling and since the outgoing pulse has an angle of 60° relative to the incoming pulse, the outgoing pulse has an angle of 30° relative to the ceiling that is between point B and the front of the spaceship (90° - 60° = 30°).

By the pulse traveling from point B on the ceiling at an angle of 30 degrees relative to the ceiling, a protractor reveals that the pulse travels about 17.5 feet in order to reach the photoelectric cell platform that is 9 feet below the ceiling. Thus, by the pulse reaching the platform, one of the photoelectric cells that cover every part of the platform is hit by the pulse. The photocell that is hit sends a current to the explosives under the platform, and the explosives ignite. Hence, with respect to the observers in spaceship Y, spaceship X is destroyed.

However, as previously pointed out, the pulse travels diagonally from A to B and diagonally from B to A and never touches the platform with respect to the occupants of spaceship X. Hence, with respect to the occupants of spaceship X, spaceship X is not destroyed and continues its flight.

Thus, as shown by the experiment, the claims and principles of special relativity bring about the impossible result of spaceship X being destroyed relative to the observers in spaceship Y while remaining whole relative to the occupants of spaceship X. Thus, by the claims of special relativity bringing about results that cannot happen, these claims are revealing their invalidity.

CHAPTER 9

No Verification of Special Relativity By the Hafele-Keating Experiment

Introduction

In October 1971, Joseph C. Hafele and Richard E. Keating conducted an experiment with the stated purpose of testing Einstein's theories of special and general relativity. The experiment involved cesium atomic clocks in the US Naval Observatory at Washington, DC, and the placing of 4 atomic clocks from the observatory on jet airplanes in regularly scheduled flights out of Washington, DC. The planes then flew the clocks around the earth twice, first eastward and then westward. The scientists predicted that during these flights and in accord with the claims of special and general relativity, the amount of time that passed on the eastward clocks would be greater by a certain amount than the amount that passed on atomic clocks in the US Naval Observatory and also that the amount of time that passed on the westward clocks would be less by a certain amount than the amount that passed on the observatory clocks. Although these differences would be extremely small because of the slow speed of the planes in comparison to the speed of light, calculations told the scientists that the atomic clocks used in their experiment would be able to reveal these differences to them. It was then claimed by the scientists that the result that they had predicted would support the claims of special relativity and general relativity.

When the westward and eastward clocks were at rest with the observatory clocks after their flights and all of the clocks were visually and physically examined by Hafele and Keating, the scientists reported that the results that they had predicted are the results that took place. Thus, according to Hafele and Keating, the results of their experiment brought verification to the claims of general and special relativity. Since the concern of this book is special relativity, it is the purpose of this chapter to show that the Hafele-Keating experiment not only fails to bring support to special relativity but also opposes what is claimed by this theory.

A report on the Hafele-Keating experiment can be found in the *November 16–18,1971 Proceedings of the Third Annual Department of Defense Precise Time and Time Interval (PTTI) Strategic Planning Meeting* in an article by Hafele titled, "Performance and Results of Portable Clocks in Aircraft." Two articles about the experiment also appeared in the July 14, 1972 edition of *Science* magazine. In the first article, "Around-the-world Atomic Clocks: Predicted Relativistic Time Gains," Hafele and Keating presented the clock rate changes that they thought would occur because of effects claimed by general and special relativity. Although this article gave anticipated results, it was actually written after the test because its predictions required information that was not available until the flights were completed. This information included the total distance and time of travel, the altitude of the airplanes, and the performance of the clocks. In the second article, "Around-the-World Atomic Clocks: Observed Relativistic Time Gains," Hafele and Keating gave results of the experiment by analyzing, interpreting, and averaging their data.

Overview of the Hafele-Keating Experiment

In taking a closer look at the claims and procedures of the Hafele and Keating experiment, we begin by turning to the "Around-the-world Atomic Clocks: Predicted Relativistic Time Gains" article. In that article, Hafele and Keating place a theoretical observer above the North Pole who does not rotate with the earth. In looking down upon the planet, he sees an assemblage of clocks on the surface of the earth at the equator. These clocks represent the atomic clocks in the US Naval Observatory. Because of earth's rotation, the North Pole observer watches the clocks move in an eastward direction at a speed of about 1,040-miles-per-hour. The observer then sees the eastbound plane and its clocks, which, we'll say, are traveling at a speed of 500-miles-per-hour. However, to the observer above the North Pole, the plane and its clocks are moving at a speed of 500-hundred-miles-per hour plus the 1,040-miles-per-hour speed of the earth's rotation for a total speed of 1,540-miles-per-hour. We'll assign a similar speed of 500-miles-per-hour to the westward plane and its clocks as they fly westward over the earth. However, the North Pole observer sees that these clocks are also traveling in an easterly direction along with earth's 1,040-mile-per-hour rotation, but their speed is reduced by their 500-hundred-miles-per hour westerly motion. This gives the westward clocks a total easterly speed of 540-miles-per-hour. Hence, according to the North Pole observer, all three sets of clocks are moving eastward at different speeds. Those on the eastward plane are the fastest, moving at 1,540-miles-per-hour. Those on the equator moving with the rotation and representing the observatory's clocks are second in speed, traveling at 1,040-miles-per-hour. And those on the westward plane are the slowest, being transported at a speed of 540-miles-per-hour.

Hence, in accord with the claims of special relativity and relative to the North Pole observer, Hafele and Keating predicted that the rate of the clocks on the eastward plane would be slower than the rate of the clocks in the Us Naval Observatory because the speed of the eastward clocks (1,540-miles-per-hour) is faster than the speed of the observatory clocks (1,040-miles-per-hour) by 500-miles-per-hour. On the other hand, the rate of the clocks on the westward plane would be faster than the rate of the clocks in the US Naval Observatory because the speed of the westward clocks (540-miles-per-hour) is slower than the speed of the observatory clocks by 500-miles-per-hour.

In addition to predicting results that are caused by special relativity, Hafele and Keating also predicted that, relative to the North Pole observer and in accord with general relativity, the altitude of the eastward and westward planes during their around-the-earth journeys and a gravitation factor would cause the rate of time in both planes and on the clocks in both planes to be faster than the rate of time in the US Naval Observatory and on the observatory clocks.

Hence, by putting together results caused by special and general relativity and by considering other factors such as the speed of the two planes in their around-the-earth journey and the amount of time these journeys would take, Hafele and Keating predicted that, relative to the North Pole observer, the journey of the clocks in the eastward plane would cause the eastward clocks to lag behind the observatory clocks by 40±23 nanoseconds. This result would be due to the claims of special relativity causing the clocks on the plane to lag 184±18 nanoseconds behind the observatory clocks and due to the claims of general relativity causing the clocks on the plane to run ahead of those in the observatory by 144±14 nanoseconds (184 - 144 = 40). Concerning the clocks in the westward plane, Hafele and Keating predicted that, relative to the North Pole observer, their journey around the earth would cause them to run 275±21 nanoseconds ahead of the observatory clocks. This result would be due to the claims of special relativity causing the clocks on the westward plane to run ahead of the observatory clocks by 96±10 and due to the claims general relativity causing the clocks on the plane to run ahead of those in the observatory by 179±18 nanoseconds (96 + 179 = 275).

In their second article, "Around-the-World Atomic Clocks: Observed Relativistic Time Gains," Hafele and Keating gave the final results of their experiment when all the moving clocks were again at rest with those in the US Naval Observatory. By actually seeing and examining the clocks from the planes and those in the observatory, Hafele and Keating determined that the eastward clocks fell 59±10 nanoseconds behind the observatory clocks and the westward clocks ran 273±7 nanoseconds ahead of those in the observatory. Hence, with the prediction that the eastward clocks would lag 40±23 nanoseconds behind the observatory clocks and with the determination that the eastward clocks lagged 59 ± 10 nanoseconds behind the observatory clocks, results concerning the eastward clocks were as predicted. In addition, with the prediction that the westward clocks would run 275±21 nanoseconds ahead of the observatory clocks and with the determination that the westward clocks ran 273±7 nanoseconds ahead of the observatory clocks, results concerning the eastward clocks were also as predicted. Hence, Hafele and Keating claimed the success of their experiment because its observed results fell within the predicted results. Thus, according to Hafele and Keating, the success of their experiment meant that the claims of special and general relative passed the test of their experiment, and hence, this meant that their experiment brought verification to the claims of these theories.

Why the Hafele-Keating Experiment Fails to Verify Special Relativity

Reason One

In what follows, an experiment will be presented that we will call the H-K2 experiment. It will take place on the basis that the Hafele-Keating experiment was conducted in accord with the principles of special relativity. Experiment H-K2 will then follow the principles as did the Hafele-Keating experiment but apply them in a different way. Experiment H-K2 will then show why the Hafele-Keating experiment fails to verify the claims and principles of special relativity.

Experiment H-K2 begins as does the Hafele-Keating experiment with the placement of an observer. However, instead of this observer being placed above the North Pole as was the case with Hafele-Keating experiment, the H-K2 observer is placed inside the eastward plane. Thus, with respect to the observer in the eastward plane and in accord with the principles of special relativity, there is a speed of 500-miles-per-hour between the observatory and the eastward plane, and this motion is entirely the motion of the observatory. This 500-miles-per-hour motion of the observatory relative to the observer in the eastward plane and the claims of special relativity cause the observatory clocks to run slower than the clocks on the eastward plane (or cause the clocks in the eastward plane to run faster than the observatory clocks).

Also with respect to the eastward plane observer, there is a speed of 1,000-miles-per-hour between the westward plane and the eastward plane, and this motion is entirely the motion of the westward plane. Furthermore, by the westward clocks having a speed of 1,000-miles-per-hour and the observatory clocks having a speed of 500-miles-per-hour with respect to the eastward plane observer, this means that the westward clocks are running slower than the observatory clocks.

Thus, with respect to the observer in the eastward plane and in accord with the claims of special relativity, the clocks in the eastward plane are running faster than the observatory clocks, and the clocks on the westward plane are running slower than the observatory clocks. Hence, relative to the eastward plane observer and in accord with the claims of special relativity, when the eastward and the westward clocks are again at rest with the observatory clocks, the eastward clocks must run ahead of the observatory clocks, and the westward clocks must lag behind the observatory clocks. However, it was reported by Hafele and Keating that when they physically and visually examined the clocks in the Hafele-Keating experiment, they determined that the claims of special relativity caused the eastward clocks to lag about 184 nanoseconds behind the observatory clocks and the westward clocks to run ahead of the observatory clocks by about 96 nanoseconds. Hence, Hafele and Keating physically determined results that oppose results that take place in accord with the claims of special relativity with respect to the eastward plane observer.

Thus, by Hafele and Keating making a physical determination in the Hafele-Keating experiment that the eastward clocks lagged about 184 nanoseconds behind the observatory clocks and the westward clocks ran ahead of the observatory clocks by about 96 nanoseconds, this determination supported results that took place in accord with the claims of special relativity with respect to the North Pole observer. On the other hand, this determination opposed results that took place in accord with the claims of special relativity with respect to the eastward plane observer. Thus by the fixed and physical results of the Hafele-Keating experiment supporting results that took place in accord with the principles of special relative with respect to the North Pole observer and by these physical results opposing results that took place in accord with the principles of special relative with respect to the eastward plane observer, the Hafele-Keating experiment bring no verification to special relativity. Furthermore, by the claims of special relativity producing results with respect to the North Pole observer that oppose results that were produced by the claims of special relativity with respect to the eastward plane observer, the claims of special relativity are shown to be defective.

Reason Two

In this section, we conduct a common special relativity experiment that reveals the way in which the claims of special relativity bring about time dilation. The experiment begins with the westward plane at rest with the observatory and a returned pulse of light being fired in the plane that travels from point A on the floor to point B on the ceiling that is vertically above point A and then returns to point A. The elapsed time of the pulse relative to observers in the observatory is the same as the elapsed time of the pulse relative to observers in the plane. This rest period is then followed by the westward plane being in uniform motion relative to the observatory at a speed of 500-miles-per-hour and the pulse again being fired in the plane. With respect to the observers in the observatory, the motion of the westward plane and the claims of special relativity cause the pulse to form the sides of an isosceles triangle and the elapsed time of the pulse to increase over what it was during the rest period because of the constant speed of light principle not allowing the speed of the pulse's source to be added to the pulse. However, with respect to observers in the westward plane, there is no increase in the elapsed time of the pulse from what it was because, in accord with the relativity principle, there is no motion of the plane to cause such an increase. Hence, by the pulse in the westward plane having an elapsed time relative to the observers in the plane that is less than the elapsed time of the pulse relative to observers in the observatory, this means that the motion of the westward plane and the claims of special relativity are causing time in the westward plane and on its clocks to pass slower than time in the observatory and on the observatory clocks with respect to observers in the observatory. This in turn means that, relative to personnel in the observatory and in accord with the claims of special relativity, less time passes on the clocks in the westward plane than on the clocks in the observatory during the motion between the westward plane and the observatory.

However, according to the Hafele-Keating, the amount of time that passed on the westward clocks during the motion between the westward plane and the observatory is greater than the amount of time that passed on the observatory clocks. Hence, instead of the Hafele-Keating experiment supporting what is claimed by special relativity, it is opposing what is claimed by the theory.

Uncertainties Concerning the Claims of the Hafele-Keating Experiment

A problem about the use of the Hafele-Keating experiment as evidence for or against special relativity involves uncertainties about the experiment and its claims. One concern centers on the fact that the radioactive material that runs the atomic clocks used in the Hafele-Keating experiment can cause the clocks to experience rate of time changes that are both random and spontaneous. Hafele and Keating state that it is these unpredictable rate changes that "produce the major uncertainties in our results" (see "Observed Relativistic Time Gains"). According to Table III in the "Performance and Results of Portable Clocks in Aircraft" article, spontaneous rate changes from a low of 1.25-nanoseconds-per-hour to a high of 5.00-nanoseconds-per-hour occurred on all 4 of the flight clocks during both the eastward and westward journeys. To understand the uncertainties that this brings to the Hafele-Keating experiment, it must first be pointed out that determinations of clock rate changes during the flight of each of the planes were made from clock readings that were taken immediately before and after the flight and not while the flight was in progress. With this in mind, we turn to clock 408 in the eastward flight and the fact that before the flight (as shown in "Performance and Results of Portable Clocks in Aircraft"), its rate was 1.78-nanoseconds-per-hour slower than the mean rate of the observatory clocks. However, at the conclusion of the eastward flight, its rate was 3.22-nanosecond-per-hour faster than the mean rate of the observatory clocks. Hence, if this rate change occurred at the beginning of the 65.42 hour eastern flight, clock 408 would have been running ahead of the observatory clocks during this flight by 3.22-nanosecond-per-hour. Hence, when these 65.42 hours ended, clock 408 would be running ahead of the observatory clocks by about 211 nanoseconds (3.22-nanosecond-per-hour x 65.42 hours = 211 nanoseconds). On the other hand, if this rate change occurred at the end of the 65.42 hour eastern flight, clock 408 would have been running behind the observatory clocks during this flight by 1.78-nanoseconds-per-hour. Hence, when these 65.42 hours ended, clock 408 would be running behind the observatory clocks by about 116 nanoseconds (1.78-nanoseconds-per-hour x 65.42 hours = 116 nanoseconds)

Hence, the spontaneous rate change of clock 408 in the eastward flight could have meant that at the end of the flight, clock 408 ran 211 nanoseconds ahead of the observatory clocks, 116 nanoseconds behind the observatory clocks, or anything in between. However, since Hafele and Keating did not know when this spontaneous change of clock 408 occurred, they did not know how many nanoseconds were either gained or lost. Hence, it was necessary for Hafele and Keating to make arbitrary decisions concerning the rate changes on clock 408 as well the rate changes that took place on the other clocks on the eastward and westward flights. These arbitrary decisions may or may not have been close to being correct, and hence, the claims made by Hafele and Keating concerning what is revealed by the clocks in their experiment are clearly uncertain.

Another concern regarding the experiment is the way in which data about clocks performance was used and interpreted. As reported in Hafele's "Performance and Results of Portable Clocks in Aircraft," the Hafele-Keating experiment was performed with the full support of the US Naval Observatory Time Service Division and with financial support from the Office of Naval Research. Official results of the experiment were given in "Performance and Results of Portable Clocks in Aircraft," an article published by the US Naval Observatory and obtainable from the US Naval Observatory Library. However, the observatory article does not give the data on which claims about clocks performance were based. Furthermore, this data was not available from the observatory library for use in this book. A specific reason for this concern about data usage can again be found and exemplified with respect to clock 408 and information given about this clock in "Performance and Results of Portable Clocks in Aircraft." In Figure 4 of this article, clock 408 is shown to have gained a few nanoseconds on the eastward flight while the other three clocks had a loss in nanoseconds. However, when articles about the experiment appeared ten months later in *Science* magazine, clock 408 appears in a column with the other three clocks and shows a loss of 55 nanoseconds, harmonizing well with the other clocks and their losses of 57, 74, and 51 nanoseconds. Although an analysis of their data undoubtedly led Hafele and Keating to give this revised statistic, a change of this magnitude and significance certainly points to the possibility that arbitrary decisions and assumptions were again involved in the use of their data.

In what follows, we share concerns about the reliability of the Hafele-Keating experiment as expressed by Louis Essen, the inventor of atomic clocks and an ardent opponent of special relativity. In "Relativity—Joke or Swindle?" an article in *Electronics & Wireless World* (Feb. 1988), Essen insists that the atomic clocks used in the Hafele-Keating experiment "were not sufficiently accurate to detect the small effect predicted." And in an earlier article titled "Relativity and Time Signals," Essen makes this statement about the experiment: "Since the difference between individual clocks was as much as 300 nanoseconds, little, if any, significance can be attached to these average values. The authors do not use all the results and apply a statistical analysis, details of which are not given, to those they do use" (*Wireless World*, Oct. 1978).

CHAPTER 10

No Verification of Special Relativity By the Frisch and Smith Muon Experiment

Among the most prominent experiments used as verification for the claims of special relativity is the 1962 experiment by D. H. Frisch and J. H. Smith, which is reported by the scientists in "Measurements of the Relativistic Time Dilation Using μ-Mesons" (*American Journal of Physics*, 1963). The experiment dealt with muons (also called μ-mesons or mu-mesons), radioactive particles that are produced in earth's atmosphere by the bombardment of incoming cosmic rays. It was the intention of Frisch and Smith to provide physical evidence for the claims and principles of special relativity through these muons. They would do this by showing how muons were able to survive a journey that required far more time than their mean lifespan would allow because of the claims and principles of special relativity functioning through the motion of the muons. It is the purpose of this article to show why the procedure, claims, and calculation of the Frisch and Smith experiment provides no evidence for special relativity.

Overview of the Frisch and Smith Experiment

The first phase of the Frisch and Smith experiment took place near the peak of New Hampshire's Mount Washington with a height of 6,288 feet. As part of the equipment used by the scientists, there was a layer of iron bars that was 2.5 feet thick, and this was used to stop muons traveling at a speed that was less than $.9950c$. Those going faster than this speed passed through the iron and entered a scintillator, a cylinder of four stacked polystyrene plastic disks. Muons going faster than $.9954c$ passed on through the scintillator, but those with speeds between $.9950c$ and $.9954c$ were trapped within the instrument. During a one-hour use of the Frisch and Smith equipment near the peak of Mt. Washington, 568 muons entered the scintillator and decayed within it. And by the 568 muons having a speed between $.9950c$ and $.9954c$, they had an average speed of about $.9952c$.

After being stopped and coming to rest in the scintillator, these 568 muons lived for an average of 2.2 microseconds before decaying into neutrinos, antineutrinos, and either positive or negative electrons. According to Frisch and Smith, this meant that the mean lifespan of these 568 muons was 2.2 microseconds upon reaching the peak of Mt. Washington. This mean lifespan of 2.2 microseconds in the scintillator was revealed to Frisch and Smith by traces of light appearing on film that were caused by each muon that was trapped in the scintillator. The length of each light trace revealed how long each muon lived after being stopped and before decaying. Copies of photographs with these light traces are included in the Frisch and Smith article "Measurements of the Relativistic Time Dilation Using μ-Mesons."

For the next part of the experiment, Frisch and Smith moved their scintillator and other equipment to a location in Cambridge, Massachusetts, that was just ten feet above sea level. The vertical length between where the scintillator was located near sea level and where the scintillator was located near the top of Mt. Washington was 6,255 feet. Although 568 muons with an average speed of .9952c entered and decayed in the scintillator in one hour when it was near the peak of Mt. Washington, the scientists wanted to find out how many muons with an average speed .9952c would enter and decay in the scintillator in one hour after traveling over this 6,255 foot length. However, although the muons had an average speed of .9952c near the top of Mt. Washington, this speed would be slightly slowed by the muons traveling through 6,255 feet of air. Hence, Frisch and Smith reduced the iron barrier from 2.5 feet near the peak of Mt. Washington to 1.5 feet near sea level in order to compensate for the air barrier. On this basis, Frisch and Smith reported that the muons traveled the 6,255 length from near the top of Mt. Washington to near sea level at an average speed of .9952c. The scientists then made calculations concerning the number of muons that would enter the scintillator at its new location without any involvement of special relativity. Hence, by the muons traveling over a length of 6,255 feet at a speed of .9952c, Frisch and Smith determined that it would take 6.4 microseconds for the muons to make this entire journey. And by the muons having a mean lifespan of 2.2 microseconds, the scientist determined that there should be about 27 muons to enter and decay in the scintillator during a one-hour period.

However, Frisch and Smith discovered that in one hour 412 muons completed this 6,255 foot journey and entered the scintillator rather than 27. As already pointed out, Frisch and Smith were able to make this discovery because of each muon that entered the scintillator creating a trace of light and by Frisch and Smith taking a picture of these light traces. Hence, it was an actual physical reality that 412 muons entered the scintillator. This brings us the question: How did it happen that 412 rather than 27 muons entered the scintillator in one hour?

According to Frisch and Smith, the answer came from special relativity and its time dilation. In their journey over a 6,255 foot length that ended at the scintillator near sea level, the muons with an enormous speed of about .9952c were in uniform motion with respect to observers on Mt. Washington (as well as with respect to observers on the sea shore by the scintillator and observers on the entire earth). In accord with the claims of special relativity, this motion of the muons relative to these observers caused a slowing of muon time by a factor of 9. For every 9 microseconds that would have passed before this motion, only one microsecond now passed during the motion. Hence, with respect to observers on Mt. Washington, instead of it taking the muons 6.4 microseconds to travel over a 6,255 foot length, it took the muons only 0.7 microsecond (6.4 ÷ 9 =0.7). Thus, by it taking only 0.7 microsecond for the muons to travel over a 6,255 foot length and by the muons having a mean lifespan of 2.2 microseconds, 412 muons were able to survive this journey and enter the scintillator in one hour. Hence, according to Frisch and Smith, the reason that 412 muons entered the scintillator at sea level with respect to the Mt. Washington observers was that muon time was dilated by a factor of 9 in accord with the claims of special relativity. Thus, according to Frisch and

Smith, their experiment brought support to the claims of special relativity by revealing how the time dilation of special relativity was able to explain why 412 out of 568 muons were able to travel over a 6,255 foot length rather than the 27 that would have traveled over this length without time dilation.

Why the Frisch and Smith Experiment Does Not Support Special Relativity

In the Frisch and Smith experiment, there is a length that extends from a point near sea level (say point B) to a point (point A) that is 6,255 feet above point B. According to Frisch and Smith, when 568 muons travel over this length at a speed of .9952c, only 27 of these muons should survive this journey because of the muons having a mean lifespan of only 2.2 microseconds when they begin the journey. However, Frisch and Smith discovered through their scintillator at point B that 412 out of 568 survived this trip. According to Frisch and Smith, the reason for this high survival rate came from the time dilation of special relativity. By the muons traveling at a speed of .9952c with respect to observers on Mt. Washington, the speed of their time was dilated by a factor of 9 in accord with the claims of special relativity. Hence, with respect to observers on Mt. Washington, instead of it taking the muons 6.4 microseconds to travel over a 6,255-foot length, it took the muons only 0.7 microsecond (6.4 ÷ 9 =0.7). According to Frisch and Smith, it was a nine-fold dilation of muon time in accord with the claims of special relativity that enabled 412 muons to survive a 6,255 foot journey.

We continue by pointing out that if there were no time dilation or length contraction by the claims and principles of special relativity, then the A to B length over which the 568 muons must move is 6,255 feet. And by the muons moving at a speed of .9952c, it would take the muons 6,285 nanoseconds to make this journey with respect to the Mt. Washington observers (*6,255 foot AB length* ÷ *.9952c speed = 6,285 nano-second elapsed time*). However, as reported by the Frisch and Smith experiment, muon time has dilated by a factor of 9 with respect to the Mt. Washington observers in accord with the claims of special relativity. Hence, because of this nine-fold slowing of muon time, it does not take 6,285 nanoseconds for the muons to make this A to B journey. Instead of 6,285 nano-seconds, it takes 693 nanoseconds for the muons to travel from A to B (6,285 nanosecond elapsed time ÷ 9 = 693 nano-second elapsed time). Hence, the A to B length over which the muons must move is now revealed by this equation: *A to B length = 693 nanosecond elapsed time x .9952c speed*. Thus, the A to B length is 690 feet (*693 nanosecond elapsed time x .9952 c speed = 690 foot AB length*). This means that the 6,255 foot A to B length over which the muons would have moved with-out length contraction is contracted by a factor of 9 (6255 ÷ 690 = 9) with respect to the Mt. Washington observers even as muon time is dilated by a factor of 9 with respect to the mountain observers. Hence, by the principles of special relativity causing muon time to dilate by a factor of 9 relative to the Mt. Washington observers, they must also cause the A to B length over which the muons move to contract by a factor of 9 relative to these observers. Thus, even as length contraction involves a nine-fold contraction in the length of each individual muon, it also involves a nine-fold contraction in the A to B length over which each muon moves.

However, it is claimed by Frisch and Smith that a nine-fold

dilation of muon time in accord with the claims of special relativity enabled 412 out of 568 muons to survive a journey over a 6,255 foot A to B length with respect to observers on Mt. Washington, rather than the 27 that would have survived without time dilation. But according to special relativity, a nine-fold slowing of muon time must also involve a nine-fold contraction of the A to B length (from 6,255 to 690 feet) with respect to the mountain observers. Hence, on the basis that 412 out of 568 muons surviving the A to B journey only by a nine-fold dilation of their time, far more than 412 out of 568 muons would have survived this journey by also traveling over an A to B length of 690 feet rather than 6,255 feet. However, the Frisch and Smith experiment omits the nine-fold contraction of the A to B length that must take place with respect Mt. Washington observers because of the nine-fold dilation of muon time that takes place with respect to these observers.

In order for the claims and principles of special relativity to cause 412 muons to survive their A to B journey, both the time dilation and length contraction of special relative must be a part of this cause. However, according to the Frisch and Smith experiment, the survival of 412 muons was caused only by time dilation. Thus, because special relativity's length contraction is not involved in the survival of the muons, the survival of 412 muons is not caused by special relativity. And because this 412 muon survival is not caused by special relativity, this survival of 412 muons brings no support or verification to special relativity.

Thus, because the Frisch and Smith experiment involves the time dilation of special relativity and omits the equally important length contraction, the experiment along with its claims and calculations is fatally flawed.

Assumptions Used in the Frisch and Smith Experiment

In the article written by Frisch and Smith concerning their experiment, "Measurement of the Relativistic Time Dilation Using μ-Mesons," sections 1, 2, and 3 deal with various aspects of the experiment. Section 4 is titled "Discussion" and begins with these words: "In this section, we discuss the major assumptions used in this experiment." Then over six pages, Frisch and Smith discuss numerous aspects of their experiment that were assumptions.

As an example of these assumptions, one of them concerned the creation of muons in earth's atmosphere. According to Frisch and Smith, it is "highly improbable that there is an appreciable number of interactions in the 0-6000 foot altitude range which create μ-mesons of the energies we would observe." As to their evidence for this claim, Frisch and Smith write, "We rely on the results of other experiments." However, no information is given about these other experiments or what they actually revealed. The reason that this is an important assumption is that their experiment dealt with a group of 568 muons that traveled to the scintillator near sea level in one hour from a point in the atmosphere that was about 6,000 feet above the scintillator. Their objective was to see how many of these 568 muons would survive this journey. However, if other muons were being created in this six-thousand-foot-high atmosphere, some of these muons could also enter the scintillator. Hence, when Frisch and Smith determined that 412 muons entered the scintillator, it could be that some of these 412 muons were part of the 568 group and that some of the muons were those created in the six-thousand-foot-high atmosphere. Thus, the claim that 412 out of the 568 group entered the scintillator would be erroneous. And by this all-important claim being erroneous, the entire experiment would be erroneous and meaningless.

Thus, in a forthright manner, Frisch and Smith pointed to various other aspects of their experiment that were assumptions and thus were uncertain. Hence, if the Frisch and Smith experiment did indeed support special relativity, this support would not be based on evidence that was about as certain as evidence can be but on evidence that was admittedly assumed and uncertain.

CHAPTER 11

No Verification of Special Relativity By the Global Positioning System

Introduction

The global positioning system enables a GPS receiver to determine where it is located on the earth. The GPS does this by having a constellation of 24 or more satellites about 12,540 miles above the planet. Each of the satellites circles the earth in about 11.96 hours by moving at a fairly constant speed of 8,724-miles-per-hour, and each satellite always passes over the earth on the same pathway. The satellites are arraigned so that if it were possible for a person to see the satellites from the earth, at least 4 of these satellites could be seen at all times. The satellites are continuously broadcasting radio signals, and each signal contains specific information. When a GPS receiver is turned on, it receives information from at least 4 of these satellites unless one or more of the signals is obstructed. This information enables a receiver to determine its location through a geometric process called trilateration. An important part of the information that comes to a receiver from the satellites is the precise time at which each signal left the satellites. This precise time, in turn, comes from one or more extremely accurate atomic clocks in each satellite. Thus, the satellite clocks and a receiver clock enable a receiver to determine the amount of time that it took for each of the signals to reach the receiver. And by determining the elapsed time of each signal from each satellite, the receiver is able to determine the length between the receiver and each of these satellites in accord with the constant speed of light. Thus, by determining the lengths between the receiver and the satellites, triangulation reveals the location of the receiver.

An early problem for the GPS concerned the GPS clocks in the satellites. When the clocks in a satellite were at rest on earth with earthbound clocks, the rate of the satellite clocks was the same as the rate of the earthbound clocks. However, when the satellite clocks were placed into an orbit around the earth about 12,540 miles above earth's surface at a fairly constant speed of 8,724-miles-per-hour, the rate of the satellite clocks was faster than the rate of the earthbound clocks by 39,000-nanoseconds-per-day. Although this is a small amount of time when we realize that a second contains a billion nanoseconds, it would have been enough to cause errors in global positioning to rapidly accumulate. The entire system would have lost its value in a very short time. However, the GPS solved this problem by slowing the rate of the satellite clocks by 39,000-nanoseconds-per-day before launching the satellite from the earth. Hence, when the satellite clocks were placed into orbit, their rate was the same as the rate of the earthbound clocks. Furthermore, if there were observers with the earthbound clocks and with the satellite clocks, the satellite clocks running slower than the earthbound clocks by 39,000-nanoseconds-per-day and the speeding up of these clocks by 39,000-nanoseconds-per-day would take place with respect to all of these observers.

The Claim that the GPS Supports the Claims of General and Special Relativity

As to why the rate of the satellite clocks was faster than the rate of the earthbound clocks by 39,000-nanoseconds-per-day relative to observers with both sets of clocks, relativity scientists claim in various articles that this phenomenon can be explained by special and general relativity. And by special and general relativity being able to explain the phenomenon, this means that the phenomenon is in accord with the claims of the two theories, and hence, the phenomenon supports the theories. One such article that is frequently quoted is "Relativity and the GPS" by Neil Ashby, professor of physics at the University of Colorado, and it is found in *Physics Today* (March 11, 2003).

According to the "Relativity and the GPS" article, the gravitational attraction between the clocks in a satellite and the earth is reduced from what it would be if the clocks were on earth by the satellite clocks being placed 12,540 miles above the earth. In accord with general relativity, this reduced gravitational attraction between the satellite and the earth causes the rate of the satellite clocks to be faster than when on the earth and faster than the rate of earthbound clocks relative to observers with both sets of clocks. It is also claimed by the "Relativity and the GPS" article that according to calculations by relativity scientists, the gravitational factor of general relativity causes the rate of the satellite clocks to be faster than when on the earth by about 46,200-nanoseconds-per-day.

As to why the rate of the GPS clocks in the satellite are faster than when on the earth and faster than the rate of the earthbound clocks by 39,000-nanoseconds-per-day rather than by 46,200-nanoseconds-per-day, this explanation comes from special relativity. According to the "Relativity and the GPS" article, the claims of special relativity and the 8,724-miles-per-hour motion between the earthbound clocks and the satellite clocks cause the rate of the satellite clocks to be slower than this rate would be without this motion by 7,200-nanoseconds-per-day relative to observers with both sets of clocks. Furthermore, since the motion between the satellite clocks and the earthbound clocks is causing no slowing of the earthbound clocks, this means that the rate of the satellite clocks is slower than the rate of the earthbound clocks by 7,200-nanoseconds-per-day with respect to observers in the satellite and observers on the earth. Hence, by the claims of general relativity concerning gravitation causing the rate of the satellite clocks to be faster than when on the earth by 46,200-nanoseconds-per-day and by the claims of special relativity causing the rate of the satellite clocks to be slower than before this motion by 7,200-nanoseconds-per-day, the net result is that the rate of the satellite clocks is faster than when on the earth and faster than the rate of the earthbound clocks by 39,000-nanoseconds-per-day with respect to observers in the satellite and on the earth. Thus, on the basis that general and special relativity can explain why the rate of the GPS clocks in the satellite are faster than when on the earth and faster than the rate of the earthbound clocks by 39,000-nanoseconds-per-day relative to observers with both sets of clocks, it is claimed that these GPS clocks bring concrete, physical verification to the claims of special and general relativity. For what is claimed by special and general relativity is what is being revealed by the GPS clocks.

Since the concern of this book is special relativity, in what follows we will be dealing only with the assertion of the "Relativity and the GPS" article that the GPS affirms the claims of special relativity.

Why the GPS Does Not Support the Claims of Special Relativity as Claimed by "Relativity and the GPS"

The experiment that follows is based on the "Relativity and the GPS" article and we will begin by placing a GPS satellite by a GPS ground station. Within each of these systems are observers as well as synchronized atomic clocks. During the first part of the experiment, we will make use of comments in the "Relativity and the GPS" article concerning twins and the "famous twin paradox" by having a twin in the satellite and the ground station. We do this in order to show that what takes place with respect to the twins is also what takes place concerning the clocks in each system. The hypothetical twins in the two at-rest systems have the same age down to a single nanosecond. Also during this rest period, we will say that GPS scientists cause no slowing of the satellite clocks before they are launched even as they cause no slowing in the aging of the satellite twin.

The Ever-Increasing Accelerating Motion of the Satellite Clocks Relative to the Ground Station Clocks

The experiment continues with the GPS scientists causing the satellite with its clocks, observers, and twin to accelerate away from the ground station at an ever-increasing speed. In order to explain what takes place during this accelerating motion concerning the satellite and the ground station clocks, the "Relativity and the GPS" article points to what takes place concerning twins in the twin paradox. As to what happens in a typical twin paradox experiment, twin A accelerates away from twin B, turns around, decelerates toward twin B, and is again at rest with twin B. Since these accelerating, turn-around, and decelerating motions are happening only with respect to twin A, the claim is made that it is during these motions that twin A ages slower than twin B with respect to both twins and observers traveling with both twins. It might be that between the accelerating, decelerating, and turn-around motions, there is uniform motion between the twins. However, during this uniform motion, the slowing that take place in the aging of twin A with respect to twin B also takes place in the aging of twin B with respect to twin A, and hence, the twin are aging at the same rate. Thus, according to the "Relativity and the GPS" article, the accelerating motion of the satellite away from the ground station at an ever-increasing speed and the principles of special relativity cause the satellite clocks to run slower than the ground station clocks with respect to all personnel in both the ground station and the satellite even as the aging of twin A in the satellite is slower than the aging of twin B in the ground station with respect to all personnel in both systems.

The Ever-Increasing Accelerating Speed Changing to a Fairly Constant Speed

A point of time then comes when the ever-increasing accelerating speed of the satellite relative to the ground station has ended and the satellite is on its assigned pathway around the earth at a fairly constant speed of 8,724-miles-per-hour. Hence, at this point and in accord with the claims of the "Relativity and the GPS" article, the principles of special relativity and the ever-increasing accelerating speed of the satellite have caused less time to pass for the satellite twin than for the ground station twin with respect to observers in the satellite and the ground station. Hence, the age of the satellite twin is less than the age of the ground station twin with respect to both sets of observers. Also at this point and in accord with the claims of the "Relativity and the GPS" article, the principles of special relativity and the ever-increasing accelerating speed of the satellite have caused less time to pass on the satellite clocks than on the ground station clocks with respect to observers in the satellite and the ground station. Hence, the time displayed on the satellite clocks is less than the time displayed on the ground station clocks with respect to both sets of observers.

However, as mentioned above, the point has come when the ever-increasing accelerating speed of the satellite relative to the ground station has ended and the satellite is now on a pathway around the earth at a fairly constant speed of 8,724-miles-per-hour. Hence, we now have before us this question: What impact does the motion between the satellite and the ground station have on the satellite clocks with respect to observers in the satellite and the ground station by this motion having a fairly constant speed of 8,724-miles-per-hour? In asking this question, we must also consider the fact the motion between the satellite and the earth involves the satellite circling the earth. This means that the motion between the two systems is an accelerating motion because of the satellite not traveling in a constant direction. Hence, this is the full question before us: What impact does the motion between the satellite and the ground station have on the satellite clocks with respect to observers in the satellite and the ground station by this motion having a fairly constant speed of 8,724-miles-per-hour and by this motion also being accelerating motion?

An answer for this question comes from the "Relativity and the GPS" article. The article claims that the 8,724-miles-per-hour motion between the satellite and the ground station causes the rate of the satellite clocks to be slower than this rate would be without this motion by 7,200-nanoseconds-per-day with respect to observers in the satellite and the ground station. Furthermore, by this motion between the satellite and the ground station causing no slowing of the ground station clocks, this means that the satellite clocks are also running slower than the ground station clocks by 7,200-nanoseconds-per-day with respect to observers in both systems. Thus, by special relativity causing the satellite clocks to run slower than the ground station clocks by 7,200-nanoseconds-per-day with respect to observers in the satellite and the ground station, and by general relativity causing the satellite clocks to run faster than the ground station clocks by 46,200-nanoseconds-per-day with respect observers in both systems, special relativity and general relativity are showing why the clocks in the satellite run faster than the clocks in the ground station by 39,000-nanosecond-per-day with respect to both sets of observers.

Thus, in answering our question concerning the impact of the motion between the satellite and the ground station on the satellite clocks, the "Relativity and the GPS" article claims that the 8,724-miles-per-hour motion between the satellite and the ground station causes the rate of the satellite clocks to be slower than this rate would be without this motion by 7,200-nanoseconds-per-day with respect to observers in the satellite and the ground station. Moreover, since the motion between the satellite and the ground station causes no slowing of the ground station clocks, this means that the satellite clocks are also running slower than the ground station clocks by 7,200-nanoseconds-per-day with respect to observers in both systems. However, this answer to our question by the "Relativity and the GPS" article brings us to another question: Is this answer by the "Relativity and the GPS" article correct? In what follows, actual physical evidence will be presented that reveals this answer to be incorrect. This evidence comes from the null results of the Michelson-Morley experiment and the null results of Michelson-Morley type of experiments, which are far more accurate than the original experiment.

To present this evidence, we begin by pointing out that the setup of the Michelson-Morley experiment is the same as the GPS setup with which we are now dealing. In the Michelson-Morley experiment, an earthbound interferometer circles the sun, and in the GPS situation, a GPS satellite circles the earth. Hence, even as the motion of the

GPS satellite around the earth at a fairly constant speed of 8,724-miles-per-hour is accelerating motion, so also is the motion of the earth around the sun at a fairly constant speed of 67,000-miles-per-hour accelerating motion.

In the case of the Michelson-Morley experiment, it was discovered that with respect to observers on the earth, such as Michelson and Morley, the 67,000-miles-per-hour motion between the earth and the sun causes no changes on the distance and elapsed times of beams fired on the earthbound interferometer from what they would be without this motion between the two system. Furthermore, if there were clocks on the interferometer revealing the elapsed times of the interferometer's beams then, with respect to observers on the earth, the motion between the earth and the sun would cause no changes on the rate of these clocks from what that rate would be without this motion. And, on the basis that the principles of special relativity are functioning through the motion between the earthbound interferometer and the sun, it is in accord with these principles that none of this motion is the motion of the earth and the earthbound interferometer with respect to personnel on earth such as Michelson and Morley. Then, by none of this motion being the motion of the earth with respect to earthbound personnel, there is no motion of the earth that can cause any change of any kind on the earth with respect to earthbound personnel.

Hence, according to the Michelson-Morley type of experiment, the motion between a GPS satellite and the earth with respect to the satellite observers causes no slowing in the rate of the satellite clocks from what that rate would be without this motion between the two systems. However, according to the "Relativity and the GPS" article and as shown previously, the motion between the earth and a GPS satellite does cause a slowing in the rate of the satellite clocks from what that rate would be without this motion not only with respect to the ground station observers but also with respect to the observers in the satellite.

Thus, the "Relativity and the GPS" article claims that the motion between a satellite and an earthbound ground station causes the rate of the satellite clocks to be slower than this rate would be without this motion between the two systems with respect to the observers in both the satellite and the ground station. However, it is shown by the Michelson-Morley type of experiment that, with respect to observers in the satellite, the motion between the satellite and the earthbound ground station causes no slowing in the rate of the satellite clocks from what that rate would be without this motion. Hence, actual physical evidence from the Michelson-Morley type of experiment reveals that what is claimed by the "Relativity and the GPS" article is erroneous.

Thus, by the "Relativity and the GPS" article making an invalid claim concerning the rate of GPS clocks in a satellite with respect to observers in the satellite, this means that the article fails to explain why GPS clocks in a satellite run faster than GPS clocks in a ground station by 39,000-nanoseconds-per-day with respect to observers in the satellite and the ground station. Hence, the article also fails to show how special and general relativity can explain why it is that GPS clocks in a satellite run faster than GPS clocks in a ground station by 39,000-nanoseconds-per-day with respect to observers in both of these systems.

What Causes the GPS Clocks to Run Fast?

By the "Relativity and the GPS" article and its claims concerning special and general relativity failing to reveal why the rate of clocks in a satellite is faster than the rate of clocks in a ground station by 39,000-nanoseconds-per-day with respect to observers in the two systems, we ask this question: Why does this phenomenon occur?

In responding to this question, we turn to "A New Era for Atomic Clocks," an article issued by the NIST (National Institute of Standards and Technology) on January 17, 2014. It was written by Laura Ost and presented on the internet. The article speaks about the recent development of "super-accurate atomic clocks that could theoretically 'tick' for 15 billion years—the age of our entire universe—without gaining or losing a second." The article continues by pointing out how such clocks "might be used to make new types of sensors measuring quantities that have tiny effects on ticking rates, including gravity, magnetic fields, force, motion, and temperature." Hence, according to the article, the "ticking" rate of atomic clocks can be changed by a number of factors, such as gravity, magnetic fields, force, motion, and temperature, without any involvement of special or general relativity. This means that when such a factor changes the rate of an atomic clock, the atomic clock is sensing and revealing the presence of this factor. Hence, in responding to the question about why GPS clocks in the satellites run faster than GPS clocks on the earth, we simply point out that various factors could be involved, such as the effect of gravity, magnetic fields, force, motion, and temperature on the satellite clocks.

Chapter 12

No Verification of Special Relativity
By the Michelson-Morley Experiment

As presented in chapter 2, the original purpose of the 1887 Michelson-Morley experiment was to detect the ether, which, it was believed, filled all of space and was the medium through which light was able to move in waves. However, the beams on the interferometer of their experiment failed to produce the result that, according to Michelson and Morley, they should have produced by the interferometer's orbital motion relative to the sun. This result caused the scientific community to conclude that the experiment was a failure with a null result. However, after the introduction of special relativity in 1905, much of the scientific world looked upon the Michelson-Morley experiment as a verification of this theory and its two principles. It is the purpose of this chapter to show why the Michelson-Morley experiment does not verify the claims and principles of special relativity.

An Overview of the
Michelson-Morley Experiment

The ether detecting purpose of the Michelson-Morley experiment was based on several basic concepts, including (1) the theoretical existence of a static luminiferous ether that filled the entire universe and all objects in the universe; (2) the determination that the earth orbits around the sun at a fairly constant speed of 67,000-miles-per-hour; (3) the reasoning that as the earth moved through the ether because of its orbit, the earth pushed against the ether and that this caused a reactive push of the ether against the earth; (4) the calculation that, by the earth traveling into the ether at the speed of the orbit, a back-and-forth beam fired on the earth that was parallel to earth's orbital motion would have a greater slowing than a back-and-forth beam that was fired at a right angle to this motion; (5) the discovery that a pattern in the widths of dark and light bands (called interference fringes) is created when two beams of light are combined and that the band pattern can be seen by causing the combined beams to enter a telescope and looking into the instrument's opposite end or by projecting the pattern on a screen or on film; and (6) the discovery that the pattern formed by combining two beams would change in a second combining of these beams if the joined crests and troughs of each beam's waves were not the same as in the first combining.

To accomplish their mission, Michelson and Morley built an interferometer, an instrument constructed on a five-foot-square block of stone and floated on a pool of mercury. The reason for constructing the interferometer on a stone block was to eliminate vibrations and other factors that could affect the results of the experiment, and the reason for floating the block on mercury was to make it possible for the stone block to easily be rotated. On the stone, a ray of light from a single projector was split into two beams that passed over pathways of about equal length. One of the beams ran from one corner of the block, A, to the corner farthest from it, C, and the length between A and C was about 4.5 feet. We will call this 4.5 foot length the AC arm of the interferometer. Hence, the length over which a returned beam would be fired would be 9 feet. However, the length over which a returned beam would travel was increased to 36 feet by having the beam traveling between 4 mirrors at point A and 4 mirrors at point C. The other beam ran from corner B to the corner farthest from it, D, and the length between B and D was about 4.5 feet. We will call this 4.5 foot length the BD arm of the interferometer. And by 4 mirrors at points B and D reflecting the beam back and forth, the length over which the beam traveled was 36 feet. As indicated, the beam over the AB arm and the beam over the BD arm intersected at a right angle.

In conducting their experiment, Michelson and Morley slowly rotated the interferometer knowing that the moment would come when one of the arms, say the AC arm, would be parallel to the orbital motion of the earth around the sun and the BD arm would be at a right angle to this motion. The scientists had determined that in this position the ether would slow down the beam on the parallel AC arm more than the beam on the right angle BD arm. This greater slowing of the beam on the AC arm than the slowing of the beam on the BD arm meant that the ether was causing a greater increase in the elapsed time of the beam on the AC arm than in the elapsed time of the beam on the BD arm. As the scientists continued to turn the interferometer, the AC arm would change from being parallel to the orbit and start becoming vertical to the orbit, and the BD arm would change from being vertical and start becoming parallel to the orbit. This would bring the arms to the moment when both arms had the same angle relative to the motion of the interferometer, and this would cause the ether to slow the beams on both arms to the same degree and cause the same increase in their elapsed times. The continuous turning of the interferometer would soon cause the BD arm to be the one that was parallel to the orbit and the AC arm to be the one that was at the right angle. In this position the ether slows the beam on the BD arm more than on the AC arm, and hence, it causes an increase in the elapsed time of the beam on the BD arm that is greater than the increase in the elapsed time of the beam on the AC arm. Thus, by the continuous turning of the interferometer causing a continuous change in the elapsed times of these two beams, the scientists were convinced that these changes in elapsed time would have a detectable impact on the dark and light bands that were formed when the beams were recombined. Their calculations told them that there would be a 40 percent change in the width of these bands.

After the completion of their experiment, Michelson and Morley published its results in "On the Relative Motion of the Earth and the Luminiferous Ether," an article in the November 1887 edition of the *American Journal of Science.* Instead of a 40 percent fringe effect, they reported that "the actual displacement was certainly less than the twentieth part of this, and probably less than the fortieth part." This meant that although their experiment revealed a change in fringe widths, the change was far less than they predicted.

Hence, because their experiment produced a result that was nowhere close to what it was believed the ether should have caused, most of the scientific world of that day concluded that the experiment was a failure with a null result and provided no evidence for the ether's existence.

The Claimed Verification of Special Relativity By the Michelson-Morley Experiment

During the decades that followed the Michelson-Morley experiment, similar experiments were performed with various technical improvements, but all of them indicated what the Michelson-Morley experiment indicated, namely that changing the direction of beams causes no change in the distance and elapsed times of beams. Then in 2009, a seminal experiment was conducted in the Stephan Schiller lab at the Heinrich-Heine University in Dusseldorf, Germany. According to the scientists conducting the experiment, its measured results were a hundred million times more precise than the Michelson-Morley1887 measurements, and these precise results revealed that earth's orbit has no impact on earth-fired beams with respect to earthbound observers. An article about this experiment titled "Michelson-Morley Experiment is Best Yet" by Harnish Johnston can be found in the September 14, 2009 *Physics World* as well as on the internet.

Thus, as shown by the Michelson-Morley type of experiments, earth's orbital motion with respect to the sun has no impact on beams of light with respect to observers on the earth. This lack of any impact on beams of light has often been looked upon as a verification of special relativity's constant speed of light principle as well as a verification of special relativity. In *Special Relativity*, a book by A. P. French, we find this statement concerning the 1887 Michelson-Morley experiment: "It has long been regarded as one of the main experimental pillars of special relativity."

However, it will be shown in what follows that the Michelson-Morley experiment and Michelson-Morley type of experiments do not verify the theory of special relativity.

Why the Michelson–Morley Experiment Does Not Verify the Claims of Special Relativity

In this section, we continue on the basis that the principles of special relativity are functioning though the 67,000-miles-per-hour motion between the earth and the sun. Thus, in accord with these principles, the motion between the earth and the sun is entirely the motion of the sun, and none of it is the motion of the earth with respect to observers on the earth, such as Michelson and Morley. Hence, by the motion between the earth and the sun not being the motion of the earthbound interferometer with respect to Michelson and Morley, there is no motion of the interferometer that can change the elapsed times of the beams fired on the interferometer with respect to Michelson and Morley. Thus, by the interferometer physically indicating to earthbound Michelson and Morley that there are no changes in the elapsed times of the interferometer's beams from what these elapsed times would be without this motion, the Michelson-Morley experiment is supporting what is claimed by special relativity.

Also according to the principles of special relativity, the motion between the earth and the sun is entirely the motion of the earth relative to the sun and theoretical observers on the sun. However, as to what takes place concerning the earthbound interferometer and its beams with respect to the observers on the sun, the Michelson-Morley experiment tells us nothing. Hence, it could be that if all the facts were known as to what takes place concerning the earthbound interferometer with respect to the sun observers, these facts supported what is claimed by the principles of special relativity. On the other hand, it could also be that, if all the facts were known as to what takes place concerning the earthbound interferometer with respect to the sun observers, these facts opposed what is claimed by the principles of special relativity. Thus, on the basis that the Michelson-Morley experiment might oppose what is claimed by the principles of special relatively, the Michelson-Morley experiment does not and cannot bring verification to the claims of special relativity. And what is the case concerning the Michelson-Morley experiment is also the case concerning other Michelson-Morley type of experiments.

CHAPTER 13

No Verification of Special Relativity's Two Principles By the Verifications of $E=mc^2$ and $E=mc^2/\sqrt{(1-v^2/c^2)}$

In the June 1905 publication of "On the Electrodynamics of Moving Bodies," Einstein presented to the world the two basic principles of special relativity and the time dilation and length contraction that can be caused by these principles when they function through uniform motion. In September of that same year, another article by Einstein was published titled "Does the Inertia of a Body Depend Upon Its Energy Content?" This article dealt with the energy and mass of material objects and presents what is claimed by equations $E=mc^2$ and $E=mc^2/\sqrt{(1-v^2/c^2)}$. In the opening sentence of "Does the Inertia of a Body Depend Upon Its Energy Content?" Einstein writes, "The results of an electrodynamic investigation recently published by me in this journal lead to a very interesting conclusion, which will be derived here." The recent publication to which Einstein refers is "On the Electrodynamics of Moving Bodies," and hence, Einstein was making it clear that he was presenting the inertia article as a supplement to the electrodynamics publication. Thus, the theory of special relativity includes the contents of both articles. Consequently, on the basis that the equations and claims of the inertia article are valid, it might be assumed that this must mean the validity of the claims and principles of special relativity as found in the "On the Electrodynamics of Moving Bodies" article. It is the purpose of this chapter to show that a validity of the $E=mc^2$ and $E=mc^2/\sqrt{(1-v^2/c^2)}$ equations and a validity of the claims in Einstein's inertia article do not mean a validation of the principles and claims in Einstein's electrodynamics article.

"Does the Inertia of a Body Depend Upon Its Energy Content?"

In his inertia article, Einstein discussed the relationship of energy and mass in two different situations. One of these situations concerns an object that is at rest relative to its observers, such as the monument of Lincoln sitting in a chair being admired by observers. According to Einstein, the mass of this monument relative to these observers equals energy, and the amount of this energy is revealed through the equation $E=mc^2$ (the energy of an object equals the mass of the object times the constant speed of light squared). However, Einstein does not give the equation in this form but though the following statement: "If a body gives off the energy L in the form of radiation, its mass diminished by L/c^2." This could be written m (mass of a body) = L (energy) $\div c^2$ (velocity of light squared), and by multiplying both sides of the equation by c^2, we have $mc^2 = L$. In later publication Einstein changed his symbol for energy from L to E, and hence, he gave us $E=mc^2$.

The other situation regarding the relationship of energy and mass presented by Einstein concerns an object that is in motion relative to observers after having been at rest with them. An example would be the sitting Lincoln monument being placed inside an airplane and the monument then being in motion relative to observers on the earth. In this case, the monument not only has energy as described by E=mc^2 with respect to the earthbound observers, but it also has kinetic energy added to it because of the added motion. According to Einstein, the total energy of this moving monument relative to the earthbound observers can be determined through the equation E=mc^2/$\sqrt{(1-v^2/c^2)}$. With respect, however, to observers traveling with the monument and at rest with it, there is no motion of the monument and no kinetic energy added to it. For them, the energy of the monument remains at mc^2.

The Basic Claims of E=mc^2 Prior to the Introduction of Special Relativity

The basic claim of E=mc^2 is that the physical matter of an object contains an enormous amount energy. In this section, we will point out that this basic claim of E=mc^2 was considered and proposed prior to the introduction of special relativity and its principles in 1905.

As an example of these pre-1905 thoughts and claims, we turn to *Physics of the Ether*, a book by S. Tolver Preston that was published in 1875. In this book Preston claims that all of space and all material objects are filled with ether that enables objects to act upon other objects over great distances. According to Preston, this ether consists of minute particles that are constantly in motion at speeds at or beyond the speed of light and that this motion of the ether particles puts energy into all physical bodies. This energy equaled "the square of the speed of the particles," even as Einstein held that the energy of an object equaled its mass times the square of the speed of light. Preston also theorized that an enormous amount of energy can come from a small amount of an object's mass by "the reduction of mass." He claimed, for instance, that a mass of one grain had enough energy to project a weight of one hundred thousand tons to a height of nearly two miles.

Another work dealing with the energy that resides in matter is the 1900 paper "Lorentz's Theory and the Principle of Reaction" by Henri Poincaré. The contents of this paper, translated into English, can be found on the internet. In this publication Poincaré speaks about the energy of an electro-dynamics field behaving like a fictitious fluid with a mass density that equaled E/c^2. This gives us m (mass density) equals E/c^2 or $E=mc^2$. In other words, Poincaré was not only speaking about an electrodynamics field containing energy but also claiming that the amount of this energy equals mc^2.

But scientists were not just theorizing about energy residing in the matter that makes up an object. There were also scientists such as Wilhelm Roentgen, Henri Becquerel, and Pierre and Marie Curie, who, prior to 1905, were conducting research concerning the invisible rays of energy that were discharged by various elements. Although there was much that they did not know about these rays, there was a growing consensus that the energy coming out of the elements was caused by the matter that made up the elements turning into energy. In 1903, Henri Becquerel gave a speech on radioactivity upon receiving the Nobel Prize for Physics with Pierre and Marie Curie. In closing, he made these comments:

> Radioactive substances, especially radium, give off energy in all the known forms: heat, light, chemical reactions, electrical charges, γ-radiation. They seem to maintain the same state indefinitely, and the source from which they derive the energy they give off escapes us. Among the hypotheses which suggest themselves to fill the gaps left by current experiments, one of the most likely lies in supposing that the emission of energy is the result of a slow modification of the atoms of the radioactive substances. Such a modification, which the methods at our disposal are unable to bring about, could certainly release energy in sufficiently large quantities to produce the observed effects, without the changes in matter being large enough to be detectable by our methods of investigation.

Hence, actual experimentation in 1903 and earlier was revealing to scientists that objects that they themselves studied and handled contained energy and that the energy leaving these objects was probably due to a modification of the atoms that made up the objects. According to these scientists, energy residing in material objects seemed to be a basic reality of some or all material objects.

Hence, without any involvement of special relativity's claims and principles, scientists were determining and physically discovering what was later claimed by Einstein through his $E=mc^2$ equation. Energy resides in matter and the amount of energy in matter is enormous: it equals mc^2.

The Basic Claims of $E=mc^2$ and Their Validation

A basic claim of $E=mc^2$ is that a material object contains energy, and this basic claim has been validated by the various uses of nuclear energy. An example is the energy from uranium being used to produce electricity. Another basic claim of $E=mc^2$ is that a small amount of matter contains an enormous amount of energy. This basic claim of $E=mc^2$ has been confirmed by the atomic bomb. Although it probably cannot be proven by these uses of nuclear energy that the energy in the matter of an object equals mc^2 in an absolute and precise sense, the basic claims of $E=mc^2$ have received solid verification, and we will proceed on that basis.

The Basic Claims of $E=mc^2$ Do Not Validate the Principles of Special Relativity

According to $E=mc^2$, every material object that we see and handle contains energy. Such energy in a material object is simply a reality of nature even as is it a reality of nature that all living creatures contain carbon. As pointed out previously, if observers are at rest with a stone monument and are viewing it, the monument before them contains energy as described by $E=mc^2$. In order for this energy to reside in the monument, no motion between the monument and the observers is needed, and hence, no functioning of special relativity's principles through this motion is needed.

Furthermore, the basic claims of $E=mc^2$ do not require the two principles of special relativity. If light did not move at c regardless of any motion by its source and if the relativity principle did not exist, material objects would still contain energy. Hence, the validity of these basic claims does not in any way validate these principles.

The Basic Claims of $E=mc^2/\sqrt{(1-v^2/c^2)}$ Do Not Validate the Principles of Special Relativity

As claimed by Einstein through $E=mc^2$, if observers are looking at a stone monument in front of them, the mass of this object equals energy. And as also claimed by Einstein through $E=mc^2/\sqrt{(1-v^2/c^2)}$, if this monument is placed in an airplane and it goes into motion relative to the earthbound observers, kinetic or motional energy is added to the monument with respect to these observers. This increased energy of the monument, in turn, increases the mass of the monument as shown by $E=mc^2$. If E (energy) increases, then m (mass) must increase because c, the constant speed of light, cannot increase. Furthermore, the basic claim of $E=mc^2/\sqrt{(1-v^2/c^2)}$ that motion increases the energy and mass of an object has frequently been verified by particle accelerators. This verification is presented in various internet articles such as "Why does a particle's mass increase as it moves?" by Dr. Sten Odenwald, an astronomer who writes for the NASA Education and Public Outreach Program.

However, as previously shown, every material object such as the Abraham Lincoln statue in the Lincoln Monument contains energy with respect to observers at rest with the statue, a reality about matter that does not require the principles of special relativity and a reality that would exist without these principles. Hence, if the Lincoln statue were placed into a moving airplane, the statue would still contain the energy that it had when on the earth with respect to observers on the earth. However, by the statue being in motion with respect to observers on earth, this motion of the statue causes it to have kinetic energy in addition to the energy that it had before this motion. This added kinetic energy to the statue means an increase in the statue's mass in accord with $E=mc^2$.

However, this added kinetic energy and mass to the statue is not caused by the claims and principles of special relativity but caused by the statue's motion. Hence, even as the principles of special relativity were not needed in order for the Lincoln statue to have energy and mass while on earth with respect to earthbound observers, the principles of special relativity are not needed in order for the Lincoln statue to have kinetic energy and mass added to it by being in motion with respect to the earthbound observers.

Furthermore, the fact that there is no need for the principles of special relativity in order for motion to add kinetic energy and mass to an object is readily revealed by the fact that the relativity principle of special relativity functions through the uniform motion between systems and that it does not function through other motions such as motions that are rough, jerky, or fluctuating. However, if motion that is rough, jerky, or fluctuating is added to the Lincoln statue, this added motion will add kinetic energy to the monument, and this increased energy will increase the mass of the monument. Hence, we are shown that the relativity principle and the relativity principle combined with the constant speed of light principle are not involved or needed in order for the motion of a system to add kinetic energy and mass to the system. Thus, the adding of kinetic energy and mass to an object by adding motion to the object as claimed by $E=mc^2/\sqrt{(1v^2/c^2)}$ does not support or verify the combined principles of special relativity.

Chapter 14

The Invalid Claim of Special Relativity

In chapters 4 through 8, experiments were presented in which the claims of special relativity were invalidated by producing impossible results. These results included the destruction of a spaceship with respect to observers before whom the spaceship is in motion while the spaceship remains whole with respect to occupants in the spaceship. And by the claims of special relativity producing impossible results, one or more of these claims must be invalid. Hence, we ask this question: Which claim or claims of special relativity cause it to be invalid? In what follows, the claim that the relativity principle of special relativity is a valid principle will be pointed to as a likely answer to this question.

Einstein's Expanded Relativity Principle

Centuries before the Einstein era, a relativity principle was in existence that was accepted throughout the scientific world and remains accepted throughout this world. Today, it is known as the "classical relativity principle." A description of the principle by Galileo was presented in chapter 1, and we repeat it here:

Shut yourself up with some friend in the main cabin below decks on some large ship, and have with you there some flies, butterflies, and other small flying animals. Have a large bowl of water with some fish in it; hang up a bottle that empties drop by drop into a wide vessel beneath it. With the ship standing still, observe carefully how the little animals fly with equal speed to all sides of the cabin. The fish swim indifferently in all directions; the drops fall into the vessel beneath; and, in throwing something to your friend, you need throw it no more strongly in one direction than another, the distances being equal; jumping with your feet together, you pass equal spaces in every direction. When you have observed all these things carefully (though there is no doubt that when the ship is standing still everything must happen in this way), have the ship proceed with any speed you like, so long as the motion is uniform and not fluctuating this way and that. You will discover not the least change in all the effects named, nor could you tell from any of them whether the ship was moving or standing still ... The cause of all these correspondences of effects is the fact that the ship's motion is common to all the things contained in it (comments by Salviati on the Second Day).

Thus, according to these words by Galileo and the classical relativity principle, if a system such as a ship is in uniform motion relative to another system such as the ocean after the two systems have been at rest with each other, then this motion can cause no change in the way things happen in the ship from the way they happened before this motion.

In chapter 5 of his book *Relativity: The Special and the General Theory*, Einstein speaks about the exactness of the classical relativity principle "in the domain of mechanics." He goes on to say that "classical mechanics affords an insufficient foundation for the physical description of all natural phenomena." Einstein then points out in chapters 5 through 7 that in order for the relativity principle to be sufficient, it must also take into account electromagnetic waves. Hence, unlike the classical relativity principle, the relativity principle of special relativity insists that when there is uniform motion between two systems, A and B, after they have been at rest with each other, this motion cannot cause any mechanical or electromagnetic event in system A to change from what it was before this motion with respect to observers in A. In like manner, this motion cannot cause any mechanical or electromagnetic event in system B to change from what it was before this motion with respect to observers in B.

We also point out that after speaking about "the great accuracy" of the classical relativity principle "in the domain of mechanics," Einstein adds the following comment: "But that a principle of such broad generality should hold with such exactness in one domain of phenomena and should be invalid for another, is *a priori* not very probable" (*Relativity: The Special and the General Theory*). According to Einstein, it was not very probable that the classical relativity principle could be so valid and exact concerning mechanics and be invalid concerning electrodynamics. Hence, Einstein did not add electrodynamics to the classical relativity principle on the basis of strong experimental evidence. He added electrodynamics to the classical relativity principle because it seemed likely that it should be added.

Hence, by Einstein's expanded relativity principle being created on the basis of what seemed likely rather than on the basis of solid evidence, it is reasonable to point to this principle as the cause for the impossible results that are produced by the claims and principles of special relativity.

Closing Comment
by the Author

In the pages of this book, I have made my case concerning the theory of special relativity and I did my best to follow the evidence wherever it led. Regardless, however, of how compelling this case may be, it gave me tremendous satisfaction to take on an enormous challenge. Furthermore, I didn't bring this challenge to an end until I was completely convinced that I had done all that I could do in meeting the challenge. As an added bonus, the research, the learning, the discoveries, and the development of new ideas were so exhilarating throughout this project that I hated to see it come to an end.

Acknowledgments
by the Author

In addition to my use of various textbooks on special relativity and the works that are listed in the bibliography, there were numerous other publications that provided me with an array of views and claims concerning special relativity and also gave me much to think about. Hence, I would like to recognize some of the works that were especially thought-provoking, including *Light Velocity and Relativity* by Arthur S. Otis, "Rethinking Relativity" by Tom Bethel, *Newton to Einstein, The Trail of Light* by Ralph Baierlein, "The GPS and the Constant Velocity of Light" by Paul Marmet, "What is wrong with relativity?" by G. Burniston Brown, "The Special Theory of Relativity: A Critical Analysis" by Louis Essen, "Relativity and the GPS" by Ronald Hatch, *Science at the Crossroads* by Herbert Dingle, *Albert Einstein: The Incorrigible Plagiarist* by Christopher Jon Bjerkness, "Hafele & Keating Tests: Did They Prove Anything?" by A. G. Kelly, *Relativity Simply Explained* by Martin Gardner, "A New Interpretation of the Hafele-Keating Experiment" by Domina Spencer, "The Michelson and Morley 1887 Experiment and the Discovery of Absolute Motion" by Reginald Cahill,

"Michelson-Morley Experiments Revisited: Systematic Errors, Consistency among Different Experiments, and Compatibility with Absolute Motion" by Hector Munera, "Time Dilation: Fact or Fiction" by J. L. Gaasenbeek, "Symmetrical Experiments to Test the Clock Paradox" by Ling Jun Wang, "Earth's Revolution and Gravity" by Ken Mellendorf, "What the Global Positioning System Tells Us about Relativity" by Tom Van Flandern; "The Michelson-Morley Experiment" by Michael Fowler, "The Special Theory of Relativity" by David H. Harrison, "Fresnel, Fizeau, Hoek, Michelson-Morley, Michelson-Gale and Sagnac in Aetherless Galilean Space" by Curt Renshaw, and "Why We Believe in Special Relativity: Experimental Support for Einstein's Theory" by John S. Reid.

Bibliography

Ashby, Neil. "Relativity and GPS," *GPS World*, Vol. 4, No. 11, Nov., 1993, 42-47.

Ashby, Neil. "Relativity and the Global Positioning System," *Physics Today*, May, 2002, 41-47.

Beck, Anna, Translator. *The Collected Papers of Albert Einstein: The Swiss Years: Writings 1909-1911*, Volume 3, Princeton University Press, Princeton, New Jersey, 1993. These collected papers contain "The Theory of Relativity" a January, 1911 lecture given by Einstein in Zurich.

Brush, Stephen G., editor. "Notes on the history of the Fitzgerald-Lorentz Contraction," *Isis*, summer, 1967, Vol. 58, No. 2.

Einstein, Albert. "Does the Inertia of a Body Depend on Its Energy Content?" Found in *Einstein's Miraculous Year*, edited by John Stachel, Princeton University Press, Princeton, New Jersey, 2005, 161-164.

Einstein, Albert. "On the Electrodynamics of Moving Bodies." Found in *Einstein's Miraculous Year*, edited by John Stachel, Princeton University Press, Princeton, New Jersey, 2005, 123-159.

Einstein, Albert. *Relativity: The Special and the General Theory*, Crown Trade Paperbacks, New York, 1961.

Einstein, Albert. *The Meaning of Relativity*, Princeton University Press, Princeton, New Jersey, 1956. Contains the four Stafford Little Lectures at Princeton University in 1921.

Einstein, Albert. "The Theory of Relativity." Found in *The Collected Papers of Albert Einstein*, vol. 3, Anna Beck, translator, Princeton University Press, Princeton, New Jersey, 1993, 348-349.

Essen, Louis. "Relativity and Time Signals, "*Wireless World*, Oct., 1978.

Essen, Louis. "Relativity - Joke or Swindle?," *Electronics & Wireless World*, Feb., 1988.

Fitzgerald, George F. "The Ether and the Earth's Atmosphere," *Science*, Vol. 13, No. 328, 1889.

Fitzgerald, George F. "The Relations between Ether and Matter," *Nature*, July 19, 1900. This article also appears in *The Scientific Writings of the Late George Francis Fitzgerald*, Joseph Larmor, editor, 1902.

Frisch, David H., James H. Smith. "Measurements of the Relativistic Time Dilation Using μ Mesons," *American Journal of Physics*, Vol. 31, 1963, 342-355.

Galilei, Galileo. *Dialogue on the Great World Systems*, Chicago University Press, Chicago, Ill., 1953.

Hafele, J. C. "Performance and Results of Portable Clocks in Aircraft, "*Proceedings of The Third Annual Department of Defense Precise Time and Time Interval (PTTI) Strategic Planning Meeting*, Nov.16-18, 1971, 261-288.

Hafele, J. C., R. E. Keating. "Around-the-World Atomic Clocks: Predicted Relativistic Time Gains" & "Around-the-World Atomic Clocks: Observed Relativistic Gains," *Science*, Vol.177, pp, 166-170, July 14, 1972.

Johnston, Harnish. "Michelson-Morley Experiment is Best Yet, "*Physics World,* Sept. 14, 2009.

Larmor, Joseph, editor. *The Scientific Writings of the Late George Francis Fitzgerald,* 1902. Contains "The Ether and the Earth's Atmosphere" by Fitzgerald.

Lorentz, Hendrik A. "Michelson's Interference Experiment," Leyden, Holland,1895. A translation of this paper from French into English can be found in *Relativity Theory: Its Origins & Impact on Modern Thought,* a book edited by L. Pearce Williams.

Lorentz, Hendrik A. "Simplified Theory of Electrical and Optical Phenomena in Moving Systems," Leiden, E. J. Brill, 1895. Written in French and made available in English on the internet by wikisource.

Michelson, Albert A., Edward W. Morley. "On the Relative Motion of the Earth and the Luminiferous Ether, "*The American Journal of Science,"* Vol. 34, No. 203, Nov., 1887, 333-345.

Nobel Foundation. *Nobel Lectures, Including Presentation Speeches and Laureates' Biographies, Physics,* 1901-1921, Elsevier, Amsterdam, 1967. Contains "On Radioactivity, a New Property of Matter," the Nobel Lecture by Antoine H. Becquerel, Dec. 11, 1903.

Odenwald, Stan. "Why does a particle's mass increase at it moves?," NASA Education and Public Outreach program, Aug. 11, 2015.

Ost, Laura. "A New Era for Atomic Clocks," an article issued by the NIST on Jan. 17, 2014, posted on the internet.

Poincaré, Henri. "The Principles of Mathematical Physics," 1904. A French to English translation of this article can be found in *Relativity Theory: Its Origins & Impact on Modern Thought*, a book edited by L. Pearce Williams.

Poincaré, Henri. "Lorentz's Theory and the Principle of Reaction," Archives of the Dutch Natural Sciences 5, 1900, 252-278. A copy of this document in French and English is posted on the internet by Wikisource la bibliothéque libre.

Preston, S. Tolver. *Physics of the Ether*, E. & F. N. Spon, London, 1875.

Stachel, John. *Einstein's Miraculous Year: Five Papers That Changed the Face of Physics*, Princeton University Press, Princeton, New Jersey, 1998. Contains Einstein's "On the Electrodynamics of Moving Bodies" and "Does the Inertia of a Body Depend on Its Energy Content?" English translations of the original German texts published in *The Swiss Years: Writings, 1900-1909*, volume 2, of the *Collected Papers of Albert Einstein*.

Williams, L. Pearce, editor. *Relativity Theory: Its Origins & Impact on Modern Thought*, John Wiley & Sons, Inc., New York, 1968. Contains "Michelson's Interference Experiment" by Hendrik A. Lorentz, 1895, and an abridgement of "The Principles of Mathematical Physics," by Henri Poincaré, 1904.